T0092528

HAZARDS ASSOCIATED
WITH HUMAN INDUCED
EXTERNAL EVENTS
IN SITE EVALUATION FOR
NUCLEAR INSTALLATIONS

The following States are Members of the International Atomic Energy Agency:

AFGHANISTAN	GERMANY	PALAU
ALBANIA	GHANA	PANAMA
ALGERIA	GREECE	PAPUA NEW GUINEA
ANGOLA	GRENADA	PARAGUAY
ANTIGUA AND BARBUDA	GUATEMALA	PERU
ARGENTINA	GUYANA	PHILIPPINES
ARMENIA	HAITI	POLAND
AUSTRALIA	HOLY SEE	PORTUGAL
AUSTRIA	HONDURAS	QATAR
AZERBAIJAN	HUNGARY	REPUBLIC OF MOLDOVA
BAHAMAS	ICELAND	ROMANIA
BAHRAIN	INDIA	RUSSIAN FEDERATION
BANGLADESH	INDONESIA	RWANDA
BARBADOS	IRAN, ISLAMIC REPUBLIC OF	SAINT KITTS AND NEVIS
BELARUS	IRAQ	SAINT LUCIA
BELGIUM	IRELAND	SAINT VINCENT AND
BELIZE	ISRAEL	THE GRENADINES
BENIN	ITALY	SAMOA
BOLIVIA, PLURINATIONAL	JAMAICA	SAN MARINO
STATE OF	JAPAN	SAUDI ARABIA
BOSNIA AND HERZEGOVINA	JORDAN	SENEGAL
BOTSWANA	KAZAKHSTAN	SERBIA
BRAZIL	KENYA	SEYCHELLES
BRUNEI DARUSSALAM	KOREA, REPUBLIC OF	SIERRA LEONE
BULGARIA	KUWAIT	SINGAPORE
BURKINA FASO	KYRGYZSTAN	SLOVAKIA
BURUNDI	LAO PEOPLE'S DEMOCRATIC	SLOVENIA
CAMBODIA	REPUBLIC	SOUTH AFRICA
CAMEROON	LATVIA	SPAIN
CANADA	LEBANON	SRI LANKA
CENTRAL AFRICAN	LESOTHO	SUDAN
REPUBLIC	LIBERIA	SWEDEN
CHAD	LIBYA	SWITZERLAND
CHILE	LIECHTENSTEIN	SYRIAN ARAB REPUBLIC
CHINA	LITHUANIA	TAJIKISTAN
COLOMBIA	LUXEMBOURG	THAILAND
COMOROS	MADAGASCAR	TOGO
CONGO	MALAWI	TONGA
COSTA RICA	MALAYSIA	TRINIDAD AND TOBAGO
CÔTE D'IVOIRE	MALI	TUNISIA
CROATIA	MALTA	TÜRKİYE
CUBA	MARSHALL ISLANDS	TURKMENISTAN
CYPRUS	MAURITANIA	UGANDA
CZECH REPUBLIC	MAURITIUS	UKRAINE
DEMOCRATIC REPUBLIC	MEXICO	UNITED ARAB EMIRATES
OF THE CONGO	MONACO	UNITED KINGDOM OF
DENMARK	MONGOLIA	GREAT BRITAIN AND
DJIBOUTI	MONTENEGRO	NORTHERN IRELAND
DOMINICA	MOROCCO	UNITED REPUBLIC
DOMINICAN REPUBLIC	MOZAMBIQUE	OF TANZANIA
ECUADOR	MYANMAR	UNITED STATES OF AMERICA
EGYPT	NAMIBIA	URUGUAY
EL SALVADOR	NEPAL	UZBEKISTAN
ERITREA	NETHERLANDS	VANUATU
ESTONIA	NEW ZEALAND	VENEZUELA, BOLIVARIAN
ESWATINI	NICARAGUA	REPUBLIC OF
ETHIOPIA	NIGER	VIET NAM
FIJI	NIGERIA	YEMEN
FINLAND	NORTH MACEDONIA	ZAMBIA
FRANCE	NORWAY	ZIMBABWE
GABON	OMAN	
GEORGIA	PAKISTAN	

The Agency's Statute was approved on 23 October 1956 by the Conference on the Statute of the IAEA held at United Nations Headquarters, New York; it entered into force on 29 July 1957. The Headquarters of the Agency are situated in Vienna. Its principal objective is "to accelerate and enlarge the contribution of atomic energy to peace, health and prosperity throughout the world".

IAEA SAFETY STANDARDS SERIES No. SSG-79

HAZARDS ASSOCIATED WITH HUMAN INDUCED EXTERNAL EVENTS IN SITE EVALUATION FOR NUCLEAR INSTALLATIONS

SPECIFIC SAFETY GUIDE

INTERNATIONAL ATOMIC ENERGY AGENCY
VIENNA, 2023

COPYRIGHT NOTICE

© IAEA, 2023

Printed by the IAEA in Austria
January 2023
STI/PUB/2036

IAEA Library Cataloguing in Publication Data

Names: International Atomic Energy Agency.
Title: Hazards associated with human induced external events in site evaluation for nuclear installations / International Atomic Energy Agency.
Description: Vienna : International Atomic Energy Agency, 2023. | Series: IAEA safety standards series, ISSN 1020–525X ; no. SSG-79 | Includes bibliographical references.
Identifiers: IAEAL 22-01546 | ISBN 978–92–0–144122–5 (paperback : alk. paper) | ISBN 978–92–0–143922–2 (pdf) | ISBN 978–92–0–144022–8 (epub)
Subjects: LCSH: Nuclear facilities. | Nuclear facilities — Safety measures. | Nuclear power plants — Safety measures. | Nuclear power plants — Risk assessment.
Classification: UDC 621.039.58 | STI/PUB/2036

FOREWORD

by Rafael Mariano Grossi
Director General

The IAEA's Statute authorizes it to "establish…standards of safety for protection of health and minimization of danger to life and property". These are standards that the IAEA must apply to its own operations, and that States can apply through their national regulations.

The IAEA started its safety standards programme in 1958 and there have been many developments since. As Director General, I am committed to ensuring that the IAEA maintains and improves upon this integrated, comprehensive and consistent set of up to date, user friendly and fit for purpose safety standards of high quality. Their proper application in the use of nuclear science and technology should offer a high level of protection for people and the environment across the world and provide the confidence necessary to allow for the ongoing use of nuclear technology for the benefit of all.

Safety is a national responsibility underpinned by a number of international conventions. The IAEA safety standards form a basis for these legal instruments and serve as a global reference to help parties meet their obligations. While safety standards are not legally binding on Member States, they are widely applied. They have become an indispensable reference point and a common denominator for the vast majority of Member States that have adopted these standards for use in national regulations to enhance safety in nuclear power generation, research reactors and fuel cycle facilities as well as in nuclear applications in medicine, industry, agriculture and research.

The IAEA safety standards are based on the practical experience of its Member States and produced through international consensus. The involvement of the members of the Safety Standards Committees, the Nuclear Security Guidance Committee and the Commission on Safety Standards is particularly important, and I am grateful to all those who contribute their knowledge and expertise to this endeavour.

The IAEA also uses these safety standards when it assists Member States through its review missions and advisory services. This helps Member States in the application of the standards and enables valuable experience and insight to be shared. Feedback from these missions and services, and lessons identified from events and experience in the use and application of the safety standards, are taken into account during their periodic revision.

I believe the IAEA safety standards and their application make an invaluable contribution to ensuring a high level of safety in the use of nuclear technology. I encourage all Member States to promote and apply these standards, and to work with the IAEA to uphold their quality now and in the future.

THE IAEA SAFETY STANDARDS

BACKGROUND

Radioactivity is a natural phenomenon and natural sources of radiation are features of the environment. Radiation and radioactive substances have many beneficial applications, ranging from power generation to uses in medicine, industry and agriculture. The radiation risks to workers and the public and to the environment that may arise from these applications have to be assessed and, if necessary, controlled.

Activities such as the medical uses of radiation, the operation of nuclear installations, the production, transport and use of radioactive material, and the management of radioactive waste must therefore be subject to standards of safety.

Regulating safety is a national responsibility. However, radiation risks may transcend national borders, and international cooperation serves to promote and enhance safety globally by exchanging experience and by improving capabilities to control hazards, to prevent accidents, to respond to emergencies and to mitigate any harmful consequences.

States have an obligation of diligence and duty of care, and are expected to fulfil their national and international undertakings and obligations.

International safety standards provide support for States in meeting their obligations under general principles of international law, such as those relating to environmental protection. International safety standards also promote and assure confidence in safety and facilitate international commerce and trade.

A global nuclear safety regime is in place and is being continuously improved. IAEA safety standards, which support the implementation of binding international instruments and national safety infrastructures, are a cornerstone of this global regime. The IAEA safety standards constitute a useful tool for contracting parties to assess their performance under these international conventions.

THE IAEA SAFETY STANDARDS

The status of the IAEA safety standards derives from the IAEA's Statute, which authorizes the IAEA to establish or adopt, in consultation and, where appropriate, in collaboration with the competent organs of the United Nations and with the specialized agencies concerned, standards of safety for protection of health and minimization of danger to life and property, and to provide for their application.

With a view to ensuring the protection of people and the environment from harmful effects of ionizing radiation, the IAEA safety standards establish fundamental safety principles, requirements and measures to control the radiation exposure of people and the release of radioactive material to the environment, to restrict the likelihood of events that might lead to a loss of control over a nuclear reactor core, nuclear chain reaction, radioactive source or any other source of radiation, and to mitigate the consequences of such events if they were to occur. The standards apply to facilities and activities that give rise to radiation risks, including nuclear installations, the use of radiation and radioactive sources, the transport of radioactive material and the management of radioactive waste.

Safety measures and security measures[1] have in common the aim of protecting human life and health and the environment. Safety measures and security measures must be designed and implemented in an integrated manner so that security measures do not compromise safety and safety measures do not compromise security.

The IAEA safety standards reflect an international consensus on what constitutes a high level of safety for protecting people and the environment from harmful effects of ionizing radiation. They are issued in the IAEA Safety Standards Series, which has three categories (see Fig. 1).

Safety Fundamentals

Safety Fundamentals present the fundamental safety objective and principles of protection and safety, and provide the basis for the safety requirements.

Safety Requirements

An integrated and consistent set of Safety Requirements establishes the requirements that must be met to ensure the protection of people and the environment, both now and in the future. The requirements are governed by the objective and principles of the Safety Fundamentals. If the requirements are not met, measures must be taken to reach or restore the required level of safety. The format and style of the requirements facilitate their use for the establishment, in a harmonized manner, of a national regulatory framework. Requirements, including numbered 'overarching' requirements, are expressed as 'shall' statements. Many requirements are not addressed to a specific party, the implication being that the appropriate parties are responsible for fulfilling them.

Safety Guides

Safety Guides provide recommendations and guidance on how to comply with the safety requirements, indicating an international consensus that it

---

[1]   See also publications issued in the IAEA Nuclear Security Series.

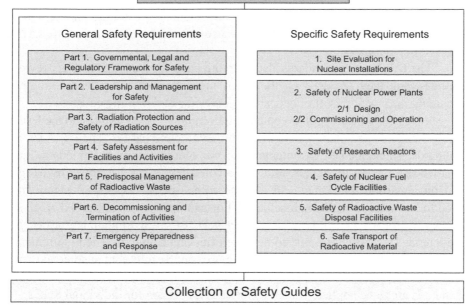

FIG. 1. The long term structure of the IAEA Safety Standards Series.

is necessary to take the measures recommended (or equivalent alternative measures). The Safety Guides present international good practices, and increasingly they reflect best practices, to help users striving to achieve high levels of safety. The recommendations provided in Safety Guides are expressed as 'should' statements.

APPLICATION OF THE IAEA SAFETY STANDARDS

The principal users of safety standards in IAEA Member States are regulatory bodies and other relevant national authorities. The IAEA safety standards are also used by co-sponsoring organizations and by many organizations that design, construct and operate nuclear facilities, as well as organizations involved in the use of radiation and radioactive sources.

The IAEA safety standards are applicable, as relevant, throughout the entire lifetime of all facilities and activities — existing and new — utilized for peaceful purposes and to protective actions to reduce existing radiation risks. They can be used by States as a reference for their national regulations in respect of facilities and activities.

The IAEA's Statute makes the safety standards binding on the IAEA in relation to its own operations and also on States in relation to IAEA assisted operations.

The IAEA safety standards also form the basis for the IAEA's safety review services, and they are used by the IAEA in support of competence building, including the development of educational curricula and training courses.

International conventions contain requirements similar to those in the IAEA safety standards and make them binding on contracting parties. The IAEA safety standards, supplemented by international conventions, industry standards and detailed national requirements, establish a consistent basis for protecting people and the environment. There will also be some special aspects of safety that need to be assessed at the national level. For example, many of the IAEA safety standards, in particular those addressing aspects of safety in planning or design, are intended to apply primarily to new facilities and activities. The requirements established in the IAEA safety standards might not be fully met at some existing facilities that were built to earlier standards. The way in which IAEA safety standards are to be applied to such facilities is a decision for individual States.

The scientific considerations underlying the IAEA safety standards provide an objective basis for decisions concerning safety; however, decision makers must also make informed judgements and must determine how best to balance the benefits of an action or an activity against the associated radiation risks and any other detrimental impacts to which it gives rise.

DEVELOPMENT PROCESS FOR THE IAEA SAFETY STANDARDS

The preparation and review of the safety standards involves the IAEA Secretariat and five Safety Standards Committees, for emergency preparedness and response (EPReSC) (as of 2016), nuclear safety (NUSSC), radiation safety (RASSC), the safety of radioactive waste (WASSC) and the safe transport of radioactive material (TRANSSC), and a Commission on Safety Standards (CSS) which oversees the IAEA safety standards programme (see Fig. 2).

All IAEA Member States may nominate experts for the Safety Standards Committees and may provide comments on draft standards. The membership of the Commission on Safety Standards is appointed by the Director General and includes senior governmental officials having responsibility for establishing national standards.

A management system has been established for the processes of planning, developing, reviewing, revising and establishing the IAEA safety standards.

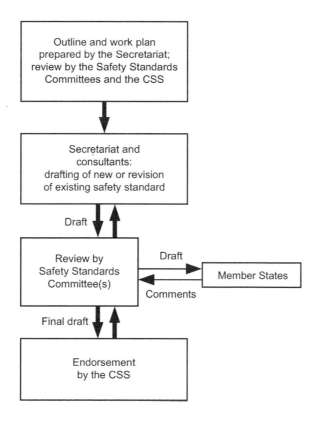

FIG. 2. The process for developing a new safety standard or revising an existing standard.

It articulates the mandate of the IAEA, the vision for the future application of the safety standards, policies and strategies, and corresponding functions and responsibilities.

INTERACTION WITH OTHER INTERNATIONAL ORGANIZATIONS

The findings of the United Nations Scientific Committee on the Effects of Atomic Radiation (UNSCEAR) and the recommendations of international expert bodies, notably the International Commission on Radiological Protection (ICRP), are taken into account in developing the IAEA safety standards. Some safety standards are developed in cooperation with other bodies in the United Nations system or other specialized agencies, including the Food and Agriculture Organization of the United Nations, the United Nations Environment Programme, the International Labour Organization, the OECD Nuclear Energy Agency, the Pan American Health Organization and the World Health Organization.

INTERPRETATION OF THE TEXT

Safety related terms are to be understood as defined in the IAEA Nuclear Safety and Security Glossary (see https://www.iaea.org/resources/publications/iaea-nuclear-safety-and-security-glossary). Otherwise, words are used with the spellings and meanings assigned to them in the latest edition of The Concise Oxford Dictionary. For Safety Guides, the English version of the text is the authoritative version.

The background and context of each standard in the IAEA Safety Standards Series and its objective, scope and structure are explained in Section 1, Introduction, of each publication.

Material for which there is no appropriate place in the body text (e.g. material that is subsidiary to or separate from the body text, is included in support of statements in the body text, or describes methods of calculation, procedures or limits and conditions) may be presented in appendices or annexes.

An appendix, if included, is considered to form an integral part of the safety standard. Material in an appendix has the same status as the body text, and the IAEA assumes authorship of it. Annexes and footnotes to the main text, if included, are used to provide practical examples or additional information or explanation. Annexes and footnotes are not integral parts of the main text. Annex material published by the IAEA is not necessarily issued under its authorship; material under other authorship may be presented in annexes to the safety standards. Extraneous material presented in annexes is excerpted and adapted as necessary to be generally useful.

CONTENTS

1. INTRODUCTION

BACKGROUND

1.1. Requirements on evaluating sites for nuclear installations[1] are established in IAEA Safety Standards Series No. SSR-1, Site Evaluation for Nuclear Installations [1]. This Safety Guide provides recommendations on how to meet the requirements established in SSR-1 [1] with regard to the evaluation of hazards associated with human induced external events[2] (HIEEs).

1.2. This Safety Guide complements other Safety Guides that provide recommendations on site evaluation and design of nuclear installations against external events excluding earthquakes [2–8].

1.3. Over the past two decades, significant new knowledge and experience have been gained in relation to hazards associated with HIEEs. This Safety Guide takes into account the following:

(a) Recent developments and regulatory requirements for assessing the safety of nuclear installations;
(b) Progress in practices in Member States relevant to hazards associated with HIEEs;
(c) A systematic approach to the identification, screening and evaluation of hazards associated with HIEEs;
(d) Good practice methodologies for evaluation of the hazards arising from the most significant HIEEs.

---

[1] The term 'nuclear installation' includes nuclear power plants; research reactors (including subcritical assemblies and critical assemblies) and any adjoining radioisotope production facilities; spent fuel storage facilities; facilities for the enrichment of uranium; nuclear fuel fabrication facilities; conversion facilities; facilities for the reprocessing of spent fuel; facilities for the predisposal management of radioactive waste arising from nuclear fuel cycle facilities; and nuclear fuel cycle related research and development facilities.

[2] In this Safety Guide, an external event is an event that originates outside the site for which the operating organization has very limited or no control over its occurrence and whose effects on the nuclear installation should be considered. Such events could be of natural or human induced origin and are identified and selected for design purposes during the site evaluation process. Events originating on the site but outside the buildings important to safety should be treated the same as off-site external events but taking into account the higher level of control over these events (this includes any coupled facilities on the site, such as those to produce hydrogen).

1.4. The terms used in this Safety Guide are to be understood as defined and explained in the IAEA Nuclear Safety and Security Glossary [9]. Explanations of technical terms specific to this Safety Guide are provided in footnotes.

1.5. This Safety Guide supersedes IAEA Safety Standards Series No. NS-G-3.1, External Human Induced Events in Site Evaluation for Nuclear Power Plants[3].

OBJECTIVE

1.6. The objective of this Safety Guide is to provide recommendations on evaluation of hazards associated with HIEEs that could affect the safety of nuclear installations, in order to meet the requirements established in SSR-1 [1], in particular Requirements 6–9, 14 and 24. These hazards need to be considered in the selection and evaluation of sites for nuclear installations, in the design of new nuclear installations and in the operation of existing nuclear installations.

1.7. This Safety Guide is intended for use by organizations involved in the identification, screening, analysis, evaluation and review of hazards associated with HIEEs, and in the provision of technical support for these activities. It is also intended for use by regulatory bodies for establishing regulatory guides on the evaluation of hazards associated with HIEEs.

SCOPE

1.8. The recommendations in this Safety Guide are intended to be used for the evaluation of hazards associated with HIEEs for nuclear installations. The approach to evaluating these hazards and the use of these evaluations need to be planned and implemented in a systematic way. This process can be divided into the following steps:

— Step 1: Identification and screening of sources of hazards;
— Step 2: Evaluation of hazards and characterization of loading conditions;
— Step 3: Design and evaluation of structures, systems and components;
— Step 4: Performance, assessment and acceptance criteria;
— Step 5: Response of the operating organization to potential HIEEs.

---

[3] INTERNATIONAL ATOMIC ENERGY AGENCY, External Human Induced Events in Site Evaluation for Nuclear Power Plants, IAEA Safety Standards Series No. NS-G-3.1, IAEA, Vienna (2002).

This Safety Guide considers steps 1 and 2. Steps 3 and 4 are addressed in IAEA Safety Standards Series No. SSG-68, Design of Nuclear Installations Against External Events Excluding Earthquakes [7], and step 5 is addressed in IAEA Safety Standards Series No. SSG-77, Protection Against Internal and External Hazards in the Operation of Nuclear Power Plants [8]. These steps are closely linked, and the needs of each step should be recognized in other steps, especially at the interfaces between steps where the outputs from earlier steps inform and provide input data to later steps.

1.9. In this Safety Guide, HIEEs are grouped into following event categories:

— External release of hazardous material;
— External explosions;
— External fire;
— Aircraft crash;
— External transport events excluding aircraft crashes;
— Other HIEEs (e.g. ground subsidence, electromagnetic interference).

1.10. This Safety Guide includes recommendations on consequential hazards arising from HIEEs, for example an aircraft fuel fire following an aircraft impact. However, it does not address combinations of hazards. Recommendations on hazard combinations are provided in SSG-68 [7].

1.11. This Safety Guide addresses a range of types of nuclear installation (see footnote 1). Many of the recommendations were originally developed for nuclear power plants, and such recommendations need to be applied to other nuclear installations through a graded approach. The direction of this graded approach is to start with recommendations relating to nuclear power plants and, if appropriate, to adjust these recommendations to installations with lesser radiological consequences. If a graded approach is not taken, the recommendations relating to nuclear power plants are to be applied.

1.12. This Safety Guide is mainly focused on the evaluation of the site for a new nuclear installation. However, the recommendations are also applicable in the re-evaluation of sites of existing nuclear installations[4] and in the periodic safety reviews of such installations. As such, the recommendations in this Safety Guide apply to all stages of the lifetime of a nuclear installation, from site selection to decommissioning.

1.13. This Safety Guide addresses site evaluation for sites on which multiple nuclear installations are located and for coupled facilities (if any) on the same site or on adjacent sites.

1.14. The external human induced events considered in this Safety Guide are of accidental origin. Other human induced events are outside the scope of this Safety Guide, although these will be a consideration in planning the mitigation of and response to such events. Considerations relating to the nuclear security of nuclear installations against malicious activities (i.e. deliberate acts of sabotage) by third parties are outside the scope of this Safety Guide. However, the methods described herein for the evaluation of hazards associated with HIEEs of accidental origin may also be applied in the evaluation of the effects of malicious acts. Guidance on nuclear security is provided in the IAEA Nuclear Security Series [10–15]. Due consideration should be given to the sensitivity of the information on HIEEs from a nuclear security perspective. Such information should be handled carefully in cooperation with nuclear security specialists.

STRUCTURE

1.15. Section 2 provides recommendations on the evaluation of hazards associated with HIEEs for nuclear installations. Section 3 provides recommendations on the identification and screening of sources of HIEEs and the evaluation of the hazards associated with these HIEEs. Section 4 provides recommendations on data collection and investigations. Sections 5–10 provide recommendations on hazard evaluations associated with the different event categories described in para. 1.9. Section 11 provides recommendations on applying a graded approach to the evaluation of hazards associated with HIEEs for nuclear installations other than

---

[4] For the purposes of this Safety Guide, existing nuclear installations are those installations that are (i) at the operational stage (including long term operation and extended temporary shutdown periods); (ii) at a pre-operational stage for which the construction of structures, the manufacturing, installation and/or assembly of components and systems, and commissioning activities are significantly advanced or fully completed; or (iii) at a temporary or permanent shutdown stage with nuclear fuel still within the facility (i.e. in the core, spent fuel pool, on-site waste storage).

nuclear power plants. Section 12 provides recommendations on the application of the management system to the evaluation of hazards associated with HIEEs. The Appendix provides tables for use in evaluating such hazards. Typical generic screening distance values are given in the Annex.

2. GENERAL RECOMMENDATIONS ON THE EVALUATION OF HUMAN INDUCED EXTERNAL EVENTS

SAFETY REQUIREMENTS FOR THE EVALUATION OF HUMAN INDUCED EXTERNAL EVENTS

2.1. Requirements 6–9, 14 and 24 of SSR-1 [1] are all relevant to the evaluation of hazards associated with HIEEs for nuclear installations, and these requirements are reproduced in paras 2.2–2.7 for convenience.

2.2. Requirement 6 of SSR-1 [1] states that "**Potential external hazards associated with natural phenomena, human induced events and human activities that could affect the region shall be identified through a screening process.**"

2.3. Requirement 7 of SSR-1 [1] states that "**The impact of natural and human induced external hazards on the safety of the nuclear installation shall be evaluated over the lifetime of the nuclear installation.**"

2.4. Requirement 8 of SSR-1 [1] states that "**If the projected design of the nuclear installation is not able to safely withstand the impact of natural and human induced external hazards, the need for site protection measures shall be evaluated.**"

2.5. Requirement 9 of SSR-1 [1] states that "**The site evaluation shall consider the potential for natural and human induced external hazards to affect multiple nuclear installations on the same site as well as on adjacent sites.**"

2.6. Requirement 14 of SSR-1 [1] states:

"The data necessary to perform an assessment of natural and human induced external hazards and to assess both the impact of the environment on the safety of the nuclear installation and the impact of the nuclear installation on people and the environment shall be collected."

2.7. Requirement 24 of SSR-1 [1] states that **"The hazards associated with human induced events on the site or in the region shall be evaluated."** Paragraphs 2.8–2.12 reproduce the supporting requirements to Requirement 24.

2.8. Paragraph 5.33 of SSR-1 [1] states:

"Human induced events to be addressed shall include, but shall not be limited to:

(a) Events associated with nearby land, river, sea or air transport (e.g. collisions and explosions);
(b) Fire, explosions, missile generation and releases of hazardous gases from industrial facilities near the site;
(c) Electromagnetic interference."

2.9. Paragraph 5.34 of SSR-1 [1] states that "Human activities that might influence the type or severity of natural hazards, such as resource extraction or other significant re-contouring of land or water or reservoir induced seismicity, shall be considered."

2.10. Paragraph 5.35 of SSR-1 [1] states that "The potential for accidental aircraft crashes on the site shall be assessed with account taken, to the extent practicable, of potential changes in future air traffic and aircraft characteristics."

2.11. Paragraph 5.36 of SSR-1 [1] states:

"Current or foreseeable activities in the region surrounding the site that involve the handling, processing, transport and/or storage of chemicals having a potential for explosions or for producing gas clouds capable of deflagration or detonation shall be addressed."

2.12. Paragraph 5.37 of SSR-1 [1] states:

"Hazards associated with chemical explosions or other releases shall be expressed in terms of heat, overpressure and toxicity (if applicable), with account taken of the effect of distance and non-favourable combinations of atmospheric conditions at the site. In addition, the potential effects of such events on site workers shall be evaluated."

2.13. The requirements equivalent to those listed in paras 2.2–2.12 for research reactors and for nuclear fuel cycle facilities are provided in IAEA Safety Standards Series Nos SSR-3, Safety of Research Reactors [16] and SSR-4, Safety of Nuclear Fuel Cycle Facilities [17], respectively.

GENERAL CONSIDERATIONS FOR THE EVALUATION OF HUMAN INDUCED EXTERNAL EVENTS

2.14. HIEEs are caused by people; the way people act creates an environment in which hazardous events can occur and propagate. An important consideration is to recognize the possibility of an event and seek data from experience to support judgements on which of these possible events are likely to be significant and on how frequently they are likely to occur. HIEEs include direct human action (e.g. exceeding a safe speed limit or energizing an incorrect item of equipment), indirect human action (e.g. substandard design of equipment, poor maintenance practice), and errors of commission and omission.

2.15. Potential sources of HIEEs are classified as stationary or mobile and both should be considered. They are defined as follows:

(a) Stationary sources of HIEEs are those that involve the handling, processing or storage of potentially hazardous substances such as explosive, flammable, corrosive, toxic or radioactive materials, and for which the location of the initiating mechanism (explosion centre, point of release of flammable or toxic gases) is fixed, such as chemical plants, oil refineries, storage depots and other nuclear facilities at the same or a nearby site. Structures such as dams that control large volumes of water are also stationary sources of HIEEs, for which recommendations are provided in IAEA Safety Standards Series No. SSG-18, Meteorological and Hydrological Hazards in Site Evaluation for Nuclear Installations [3].
(b) Mobile sources of HIEEs are those for which the location of the initiating mechanism is not totally constrained, such as the transport or movement of

hazardous material or potential projectiles (e.g. by road, rail, waterways, air, pipelines). In such cases, an accidental explosion or a release of hazardous material might occur anywhere along a road, route or pipeline.

2.16. A region with a nuclear installation site is required to be examined for facilities and human activities that have the potential to endanger the nuclear installation over its entire lifetime (see para. 4.12 of SSR-1 [1]). As such, each potential source of HIEEs is required to be identified and assessed to determine the potential interactions with the nuclear installation.

2.17. Paragraph 4.14 of SSR-1 [1] states that "The size of the region to be investigated shall be defined for each of the natural and human induced external hazards." The size of the region to be investigated depends on the type of HIEE source and will range from a few kilometres for fire to tens of kilometres for aircraft crashes and bombing ranges, for example. The possibility that, in specific situations, a minor event might lead to severe effects should be taken into account.

2.18. Some of the hazards associated with HIEEs are more widespread than others. These effects could affect the nuclear installation's off-site facilities as well as operating personnel and items important to safety on the site, such as by affecting the availability of evacuation routes (e.g. the site might lose links to safe areas in the region), the effective implementation of emergency procedures (e.g. access by operating personnel could be impaired), and the availability of the external power grid and the ultimate heat sink (see also Requirement 11 of SSR-1 [1]). Special attention should be given to understanding the various levels of defence in depth that might be challenged by such events.

2.19. Paragraph 4.15 of SSR-1 [1] states that "The site and the region shall be studied to evaluate the present and foreseeable future characteristics that could have an impact on the safety of the nuclear installation." Similarly, Requirement 10 of SSR-1 [1] states that "**The external hazards and the site characteristics shall be assessed in terms of their potential for changing over time and the potential impact of these changes shall be evaluated.**" New sources of HIEEs can appear and existing sources can evolve rapidly. Therefore, a prognosis should be made for possible regional development over the anticipated lifetime of the nuclear installation, with account taken of the degree of administrative control that could realistically be exercised over activities in the region. In this respect, allowance should be made for the fact that technologies in the chemical and petrochemical industries, as well as traffic densities, may evolve rapidly.

2.20. HIEEs initiated at a source might eventually result in different hazards at a nuclear installation site following an interacting mechanism[5]. A number of potential HIEE sources (e.g. a chemical process site) are presumed to exist around a nuclear installation; each source is capable of one or more events (e.g. a facility failure causing an explosion and releasing stored process gas), and each event might create one or more hazardous conditions (e.g. explosion pressure wave, release of toxic gas) with the potential to challenge safety at a nearby nuclear installation. In principle, it is necessary to perform a hazard analysis of each HIEE scenario; however, only a small subset of these scenarios are likely to represent a credible risk to safety. In order to make the overall HIEE analysis traceable, this Safety Guide includes recommendations on identification and screening to ensure that only those sequences that are significant to the safety of the nuclear installation are considered throughout the entire process.

2.21. In general, there are three types of protection against HIEEs for a nuclear installation: (i) protection through a robust design of the structures, systems and components important to safety; (ii) protection through the provision of site protection measures such as sufficient distance and barriers; and (iii) protection through administrative measures such as no-fly zones and restrictions on the transport of hazardous material in the vicinity of the site. Administrative measures are generally the least reliable means of protection and should be considered as complementing the first two types of protection.

2.22. Paragraph 4.19 of SSR-1 [1] states:

"External hazards that are not excluded by the screening process shall be evaluated and then used in establishing the site specific design parameters and in the re-evaluation of the site, in accordance with the significance of these hazards to the safety of the nuclear installation."

A satisfactory engineering solution should be implemented to protect against those HIEEs that have not otherwise been excluded from further consideration using the screening process presented in Section 3. Appropriate administrative actions should be taken in the case of an existing nuclear installation in which satisfactory engineering solutions are not considered reasonably practicable.

---

[5] To further illustrate the concept of 'interacting mechanism', examples of HIEE event categories, generic screening distance values, identification of sources of HIEEs, potential HIEEs at these sources, possible hazards at a site, load characterization parameters and possible consequences at a nuclear installation site are provided in the Appendix and in the Annex.

2.23. Lack of confidence in the quality of the available data (e.g. in terms of their accuracy, applicability, completeness or quantity) may preclude the use of complex analysis techniques to characterize some HIEEs, either at the screening step or in the subsequent hazard evaluation. In such cases, a pragmatic approach based on engineering judgement should be taken, always ensuring that such judgements are demonstrably conservative (see also para. 4.8 of SSR-1 [1]). Recommendations on data collection are provided in Section 4.

2.24. Hazards associated with HIEEs in the region of nuclear installations are required to be periodically re-evaluated within the framework of the periodic safety reviews of nuclear installations (see Requirement 29 of SSR-1 [1]).

3. IDENTIFICATION OF SOURCES OF HUMAN INDUCED EXTERNAL EVENTS, SCREENING AND EVALUATION METHODS

GENERAL PROCEDURE

3.1. The evaluation of hazards associated with HIEEs involves a multistep approach (see para. 1.8). In the first step, sources of HIEEs should be identified on the basis of available data, followed by collection of data for the relevant regions. Screening should then be conducted on the basis of the established distance and probability criteria. In the next step, detailed evaluation of screened-in hazards should be conducted. The identification of sources of HIEEs should initially be performed using limited, easily accessible data, and should then be refined as more data, knowledge and information regarding how the HIEEs might affect the site or nuclear installation become available. Recommendations on the process of identification, screening and detailed evaluation of each source of HIEEs are provided in this section and are shown in Fig. 1.

IDENTIFICATION AND SCREENING OF SOURCES OF HUMAN INDUCED EXTERNAL EVENTS

3.2. The screening distance value is the distance from the nuclear installation site beyond which a hazard from an HIEE is considered insignificant to the safety of the nuclear installation. The screening distance value is a simple and conservative

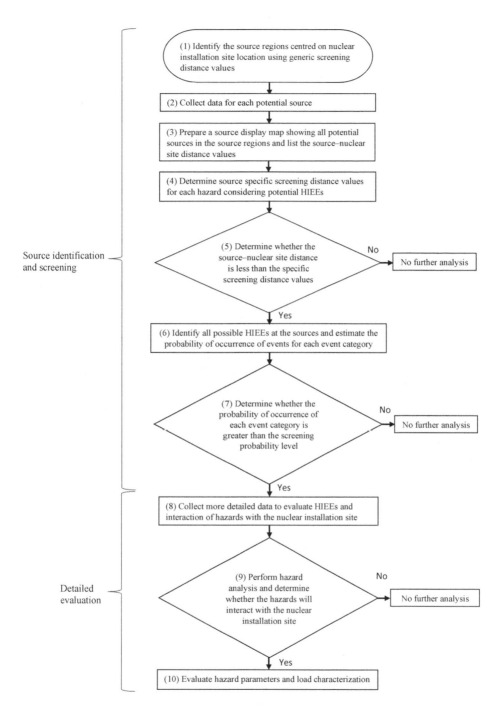

FIG. 1. *Process for source identification, screening, and detailed evaluation for each type of source of HIEEs.*

tool linked to the potential hazard that ignores any additional factors such as the mass involved or typical atmospheric conditions. For some sources, a simple deterministic study based on information on the location and characteristics of the source may be enough to show that no interaction takes place.

3.3. To initiate the evaluation process, source regions centred on the nuclear installation site should be identified (see box 1 in Fig. 1) on the basis of generic screening distance values for different event categories (see Table A–1 in the Annex). These generic screening distance values are typical values used by some States for large nuclear power plants with standardized designs. For other types of nuclear installation, these values should be reviewed and revised accordingly. These values should also be revised if the nuclear installation design and layout present any specific potential weakness with respect to HIEEs.

3.4. Local topography and regional and local meteorological effects may significantly modify the initially assumed safe distances. If there are any peculiar site conditions or significant specific hazards, the sources of HIEEs should be considered in the next evaluation step even if they were screened out in the previous evaluation step with respect to distance. Safe distances from potential sources differ greatly, for example for a chemical plant located close to a nuclear installation that is well protected by hills compared with a nuclear installation located farther away in a flat area with predominant winds blowing towards the site.

3.5. All stationary and mobile sources of potential HIEEs in the source regions should be identified, and data for these sources (e.g. source type, distance, potential events) should be collected (see box 2 in Fig. 1). Recommendations on data collection and investigations are provided in Section 4.

3.6. A source display map showing all potential sources of HIEEs (both present and foreseeable sources) should be prepared, and these sources should be listed together with the distances from the nuclear installation site (see box 3 in Fig. 1). Any uncertainties related to these sources should be estimated.

3.7. For each type of effect that could arise from an HIEE, the acceptable loading limit for the nuclear installation design should be considered.

3.8. A specific screening distance value for each source of an HIEE (stationary and mobile) should be determined by simple calculations using source specific data and considering local site conditions. The determination of the specific screening distance value should take into consideration the severity and extent of the event, including relevant uncertainties, as well as the expected characteristics

12

of the nuclear installation to be located at the site. For the early stages of the siting process, these characteristics may be assumed to be those corresponding to the standard nuclear installation design.

3.9. HIEEs might potentially generate different types of hazard (e.g. an event at a chemical plant might produce toxic gas and a pressure wave) at the nuclear installation site (see box 4 in Fig. 1), as explained in para. 2.20. The specific screening distance value of each hazard will be quite different as a gas vapour cloud may travel a much longer distance than the pressure wave. In this case, the screening distance value of this source should be taken as the longer distance.

3.10. After considering potential future changes in source characteristics (see para. 2.19) and associated uncertainties related to distances and intensities, if the nuclear installation site is beyond all specific screening distance values for the specific source of HIEEs, no further analysis is necessary (see box 5 in Fig. 1).

3.11. For sources of HIEEs that generate effects of the same nature, a further screening should be performed. This screening should be based on an enveloping criterion and should exclude those sources that generate events that are enveloped by other sources of HIEEs, even if the site is within the specific screening distance values for these sources. However, it should be ensured that the enveloped sources are considered if and when the event frequency is estimated. Care is also needed to avoid interpreting this enveloping as a reduction in the number of events that could affect the nuclear installation, and thus a reduction in the event probability.

3.12. If the nuclear installation site is within one or more specific screening distance values, relevant HIEEs are required to be identified and the probability of occurrence of these events is required to be estimated (see box 6 in Fig. 1) (see Requirement 6 of SSR-1 [1]).

3.13. The probabilistic screening should be done on the total occurrence of an event category. If the probability of occurrence is less than the specified screening probability level[6], no further analysis is necessary for that source (see box 7 in

---

[6] The screening probability level is based on the probability of the occurrence of events and is defined as the limiting value of the annual probability of occurrence of events with potential radiological consequences. In some States, a probability of 10^{-7} per reactor-year is used in the design of new facilities as one acceptable limit on the probability value for interacting events with serious radiological consequences, and this is considered a conservative value for the screening probability level if applied to all events of the same type (e.g. all aircraft crashes, all explosions).

Fig. 1). The screening probability level should be chosen such that the radiation risk associated with hazards is acceptably low. Uncertainties should be considered in calculating the probabilities of occurrence of HIEEs in probabilistic screening.

DETAILED EVALUATION OF HUMAN INDUCED EXTERNAL HAZARDS INCLUDING HAZARD PARAMETERS AND LOAD CHARACTERIZATION

3.14. If the probability of occurrence of the HIEEs under consideration is greater than the specified screening probability level, a detailed evaluation should be performed. For this purpose, more detailed data should be collected to evaluate the events and the interaction of the hazards with the nuclear installation site (see box 8 in Fig. 1).

3.15. Hazard analysis should be performed to check whether hazards associated with HIEEs will interact[7] with the nuclear installation site. If the hazard analysis results show that the hazards will not interact with the nuclear installation site, no further analysis is necessary (see box 9 in Fig. 1).

3.16. If any of the hazards can interact with the nuclear installation site, a detailed hazard evaluation should be performed and hazard parameters and load characterization should be established (see box 10 in Fig. 1). Tables 3 and 4 in the Appendix list the common hazards likely to be encountered are listed and the relevant type of hazard and characterization parameters are indicated in each case.

3.17. If applicable, a second level of screening based on the specific characteristics of the site and the nuclear installation can be implemented. Typical screening parameters to be applied are probability, magnitude and distance of the HIEE, and on-site characteristics (e.g. design conditions, zones of influence). Details are provided in Ref. [18].

3.18. This process should be repeated for each source of HIEEs. Further recommendations on the application of the process for each event category are provided in Sections 5–10.

---

[7] Interact means that a hazard will reach the nuclear installation site, as determined by hazard analysis.

4. DATA COLLECTION AND INVESTIGATIONS REGARDING HUMAN INDUCED EXTERNAL EVENTS

4.1. The collection of data regarding potential sources of HIEEs should involve the collection of site specific data as well as generic data on events due to similar sources worldwide, as such events might or might not have occurred around nuclear installation sites. It should be recognized that such data might not be readily available for reasons of confidentiality.

4.2. Individual States have different methods of data collection. The recommendations in this section provide a general approach for data and information collection that should be adapted to the specific legal framework of the State in which the nuclear installation site is situated.

DATA AND INFORMATION COLLECTION RESOURCES FOR HUMAN INDUCED EXTERNAL EVENTS

4.3. Requirements for data and information collection are established in Requirement 14 of SSR-1 [1]. The following is a list of the most relevant and important data and information collection resources:

(a) Organizations and individuals responsible for potential sources of HIEEs;
(b) Local and national government organizations with an interest in controlling, licensing or authorizing sources of HIEEs, including relevant authorities involved in the regulation of health and safety;
(c) Professional institutions and organizations;
(d) Regional data and relevant documents from government organizations, supplemented by generic data from the literature;
(e) Experience of good practice in defining hazards from similar sources that are potentially significant to nuclear installations elsewhere;
(f) Other sources of data such as local maps, published reports and public records relevant to activities around the nuclear installation site and which are likely to be relevant to HIEEs;
(g) Public and private agencies and individuals (in additional to those identified above) likely to be knowledgeable about the characteristics of the local area.

Seeking advice from organizations and individuals responsible for potential sources of human induced external events

4.4. The most important data and information resource regarding the hazards arising from a source of HIEEs is the operating organization of the source itself. Contact with the operating organization should be made at an early stage, with the objective of building a constructive relationship to facilitate information exchange. It is important to remember that while the source (e.g. an industrial site) presents a portfolio of hazards to the nuclear installation site, the nuclear installation also presents a portfolio of hazards to the source of HIEEs.

4.5. The operating organization of the source of HIEEs is likely to have the best understanding of the processes and hazards presented by its activities. The operating organization may already have well developed data and safety analyses that could be made available and almost certainly will be the best source of expert advice on its activities.

4.6. The operating organization of the source of HIEEs is likely to be subject to health and safety regulation. The appropriate regulatory bodies should be consulted for advice and should be made aware of the potential development of the nuclear installation and the likely hazards it might pose to industrial sites in the region. The operating organization of the nuclear installation should ensure that it provides a clear description of the aim and scope of the data request in order to ensure the quality and accuracy of the gathered data.

4.7. The information received from operating organizations of the sources of HIEEs should be verified and validated and, wherever possible, also be validated by an independent reviewer.

Regional emergency plans

4.8. Industrial sites that could impose hazards on a nearby nuclear installation are likely to also expose the local population to the same hazards. Such sites should be expected to provide sufficient data to enable national or local government authorities (as appropriate) to prepare regional emergency plans. Such government authorities may have useful data on regional sources of HIEEs that should also be collected.

Land use planning

4.9. Many States have well developed land use planning legislation that will apply to any new or proposed nuclear or conventional development; this same legislation is also likely to have been applied to any existing sources of HIEEs in the region at the time of their planning and development. An objective of land use planning legislation is usually to ensure that all national and local government agencies requiring knowledge of a planned hazardous site are able to obtain the information they need at an appropriate stage before and during the development process (including the data needed for the development of regional emergency plans) and have the opportunity to provide advice during the planning process on any public safety issues raised by the development. A further objective is to provide a platform for informing those members of the public (including the operating organizations of other industrial sites) who might be affected by the development and for facilitating public comment. The government planning authority for the region surrounding the nuclear installation may be able to provide useful information on sources of HIEEs. The degree to which land use planning legislation considers subsurface land use differs between States. The potential for subsurface human activities to change the external hazards for a nuclear installation should be considered under the national legal framework (see also para. 5.34 of SSR-1 [1]).

4.10. Consideration should be given to sources of HIEEs that are planned or under commercial development, watercourse developments such as dams, and marine developments such as new or modified ports and harbours (and associated changes to sea lanes) and barrages, as well as to any sources of HIEEs that are undergoing decommissioning. Such developments might lead to additional sources of hazards in the future and potentially to an increased risk of radiological consequences over the lifetime of the nuclear installation. Also important are developments that could change the population distribution in the region around the nuclear installation, since this might have implications for emergency preparedness and response.

4.11. Particular consideration should be given to the possibility that new sources of HIEEs could present hazards that are the same as hazards from existing sources that are currently screened out. The potential for adverse interactions between any new hazards and those from existing sources should also be considered (e.g. the possibility of fire spreading from a new source of HIEEs to an existing source). In either case, it may be necessary to provide additional protection and/or mitigation measures either at the nuclear installation site or as part of the new development. The progression of industrial development should be closely tracked by maintaining a continuous liaison with the local authorities.

Military sites and civil sites undertaking national defence work

4.12. Military sites and civil sites undertaking national defence work will almost always be subject to extensive restrictions on the dissemination of information about the processes and activities that take place on them, which might make it impossible for the operating organization of a nuclear installation to undertake a credible safety analysis of potential HIEEs arising from such sites. Regulatory bodies or other government agencies may have preferential access or even information exchange agreements with the defence agencies controlling these sites. Operating organizations of nuclear installations should seek advice from the regulatory body on the need for and the necessary extent of HIEE safety analysis in these cases. If specific information is not made available, generic data can be used.

DATA AND INFORMATION ON HUMAN INDUCED EXTERNAL EVENTS

4.13. Paragraph 1.9 lists six major categories of HIEE that should be considered. The region surrounding the nuclear installation site should be investigated for the presence of any human activities that have the potential to cause events in these categories. The size of the region to be investigated will depend on the nature of the human induced activities taking place. For example, the presence of a large petrochemical site storing very large quantities of hazardous material might have the potential to affect a larger geographical area in the event of an accident than, say, a small quarrying site storing and using only limited quantities of mining explosives. Table A–1 in the Annex provides generic screening distance values that are considered representative of common hazards belonging to each event category and their ability to affect a nuclear installation site.

Data uncertainty and the use of expert judgement

4.14. For many HIEEs there is often insufficient information available locally to perform a reliable evaluation of the probability of occurrence and probable severity of the event. It may therefore be useful to obtain statistical data on a national, regional or global basis. Values obtained in this way should be examined to determine whether they should be adjusted to compensate for any unusual characteristics of the source or of the nuclear installation site and the surrounding area. Where there is no reliable basis for calculating the severity of the effects of an HIEE using local data, all available information and assumptions about that event should be obtained on a global basis and the hazard evaluation should be undertaken including expert judgement.

STATIONARY SOURCES OF HUMAN INDUCED EXTERNAL EVENTS

4.15. The following information for stationary sources should be collected, although the level of detail could differ depending on the specific site conditions and the site evaluation stage:

(a) The nature of hazardous material involved and the quantities in storage, being processed on the source site or in transit in the vicinity;
(b) The types of storage and processes;
(c) The dimensions of major vessels, stores or other means of confinement;
(d) The location and distances to the nuclear installation site of these means of confinement, their construction and their isolation systems;
(e) The operating conditions of these means of confinement (including the frequency of maintenance);
(f) The active and passive safety features of these means of confinement.

4.16. The severity of the hazard might not be directly related to the size of the facilities on the source site, but the maximum amount of hazardous material present at any given time and the processes in which it is used should be taken into consideration in establishing the significance of the source to the safety of the nuclear installation site. Furthermore, the progression of an accident with time, such as fire spreading from one tank to another on the source site, should also be considered.

4.17. Pipelines carrying hazardous material from or between different stationary source locations should be considered, as mobile sources. Specific consideration should be given to industrial hydrogen storage and distribution for domestic use.

4.18. Other sources to be considered include construction yards and mines and quarries that use and store explosives.

4.19. Explosives that can generate pressure waves, projectiles and ground shock are used at mines and quarries; moreover, mining and quarrying can cause ground collapse, subsidence and landslides. Information should be obtained on the locations of all past, present and possible future mining and quarrying work and the maximum quantities of explosives that may be stored at each location. Information on geological and geophysical characteristics of the subsurface in the area should also be obtained to ensure that the nuclear installation is safe from ground collapse or landslide caused by such activities.

4.20. Fracking[8] activities and other means of natural gas extraction should also be considered, as they are similar to mining activities in that they can cause ground vibrations, subsidence and even ground failure.

4.21. At military installations, hazardous material is handled, stored and used, including in activities such as firing range practice and handling of munitions. Military airports and their associated air traffic systems, including training areas, should be considered as potential sources of HIEEs.

MOBILE SOURCES OF HUMAN INDUCED EXTERNAL EVENTS

4.22. Mobile sources of HIEEs are typically aircraft (any crewed or uncrewed aerial vehicles), road and rail vehicles, sea and river transport vessels, and pipelines. Air traffic presents a different type of mobile source of HIEEs because of the possibility of an aircraft crash directly into the nuclear installation, and this should be taken into account.

4.23. The hazards to a nuclear installation arising from surface transport (e.g. by road, rail, sea, inland waterways or pipelines) are similar to those from industrial plants. The transport and movement of hazardous material between collocated nuclear installations should also be considered, as potential sources of HIEEs.

Air transport

4.24. With regard to aircraft crash hazards (see para. 5.35 of SSR-1 [1]), a study should be made of the following:

(a) Local airports and their layout, take-off, landing and holding patterns and procedures, types of aircraft and movement frequencies.
(b) Air traffic corridors (airways) and other designated restrictions to flight transit (e.g. restricted and prohibited zones).
(c) Information on aircraft accidents for the region and for similar types of airport and air traffic. Information should be collected for general aviation

---

[8] Fracking is a process by which liquid is injected at high pressure into the ground to force open existing fissures and extract oil, natural gas, geothermal energy or water from deep underground.

and for civil and military air traffic. Of particular interest are military aircraft training areas (especially low flying areas) and areas within the region used for filling firefighting aircraft with water, since these might be areas of relatively high crash probability.

(d) Information on crash rates of each aircraft type flying near the nuclear installation in the respective flight mode (i.e. in flight, landing and taking off, including normal or special flight modes for military aircraft).

4.25. The size of the geographical region considered for aircraft crash hazard should, in general, be larger than that for other sources of HIEEs.

Transport of hazardous material by sea and inland waterways

4.26. The transport of hazardous material by sea or inland waterways might present a significant hazard. In addition to the accidental release of flammable or toxic gases and/or vapours, vessels, their loads or possible water-borne debris could block or damage cooling water intakes and outfalls associated with ultimate heat sinks. Other cargo that is not formally classified as hazardous material, such as thick liquids, pastes, absorbent bulky freight (e.g. wood pellets) and sticky chemicals, could also jeopardize cooling water intakes and outfalls associated with ultimate heat sinks.

4.27. Most sea traffic accidents occur in coastal waters or harbours; therefore, shipping lanes near the site should be identified. Information should be collected on the characteristics of shipping traffic in the region, such as the following:

(a) The location of shipping lanes local to the nuclear installation site;
(b) The nature, types and quantities of hazardous material conveyed along a route in a single transport movement;
(c) The sizes, numbers and types of vessels;
(d) The points of closest approach to the nuclear installation site;
(e) Accident statistics including consequences.

Harbours should be also studied as stationary sources of HIEEs owing to the presence of cargo containing hazardous material.

Transport of hazardous material by road and rail

4.28. Railway wagons and road vehicles, together with their loads, are potential sources of HIEEs that should be given careful attention, particularly for busy routes, junctions, marshalling yards and loading areas. Information should be collected on the characteristics of traffic flows in the region, such as the following:

(a) The location of road and rail routes local to the nuclear installation site;
(b) The nature, types and quantities of hazardous material conveyed along a route in a single transport movement;
(c) The sizes, numbers and types of vehicle;
(d) The points of closest approach to the nuclear installation site;
(e) Speed limits, control systems and safety devices;
(f) Accident statistics including consequences.

Marshalling yards should be also studied as stationary sources of HIEEs owing to the presence of cargo containing hazardous material.

Transport of hazardous material by pipeline

4.29. The following is a typical set of data and information that should be collected for pipelines:

(a) The location of pipe routes local to the nuclear installation site;
(b) Whether the pipeline is on the surface or buried near the nuclear installation site, and the diameter of the pipe;
(c) The nature of the materials transported and the flow capacity and internal pressure;
(d) The distances between valves or pumping stations;
(e) The point of closest approach to the nuclear installation site;
(f) Safety features, and relevant accident records including consequences.

SOURCE DISPLAY MAP OF HUMAN INDUCED EXTERNAL EVENTS

4.30. Source display maps should be prepared, preferably using a geographical information system (GIS) platform, showing the locations and distances from the nuclear installation of all sources of HIEEs identified in the data collection step and the size of the regions considered for each hazard type. Stationary sources and mobile sources of HIEEs should be indicated, noting transport routes close to the site, the regions considered and the most hazardous point (normally the point of closest approach) for each route. Any unusual features should be shown, such as sources of HIEEs whose hazards interact to provide an increased challenge to the safety of the nuclear installation site.

4.31. The source display maps should also reflect any foreseeable developments in human activity that might affect safety over the projected lifetime of the nuclear installation.

5. HUMAN INDUCED EXTERNAL EVENTS INVOLVING THE RELEASE OF HAZARDOUS MATERIAL

5.1. Hazardous material is normally kept in closed containers but upon release could cause a hazard to operating personnel and to items important to safety at a nuclear installation site. The following materials should be considered:

(a) Flammable gases, liquids, vapours and aerosols that can enter ventilation system intakes and burn or explode;
(b) Toxic and asphyxiant gases and liquids that can threaten human life or indirectly impair safety functions (especially gases heavier than air, such as carbon dioxide and chlorine, which can cause serious health effects);
(c) Corrosive and radioactive gases and liquids that can threaten human life or directly impair safety functions associated with structures, systems and components.

5.2. HIEEs and dispersion mechanisms are addressed in this section; explosive effects are addressed in Section 6. The ways in which these different materials affect structures, systems and components and personnel at a nuclear installation differ substantially and are covered in detail in other Safety Guides (e.g. SSG-68 [7]); however, the propagation phenomena from the source of HIEEs to the nuclear installation site are addressed in this section.

HUMAN INDUCED EXTERNAL EVENTS INVOLVING HAZARDOUS LIQUIDS

5.3. Hazardous liquids can be released on land, into water bodies and into the ground. A significant factor affecting the dispersion mechanisms for liquids is the local topography and type of soil between the source of HIEEs and the nuclear installation site. Liquids disperse across land primarily under gravity by flowing downhill; their dispersion is therefore heavily dependent on regional and source-to-site topographical features and is very likely to be directional, and this should be considered. The dispersion also depends on the roughness of the ground, which differs depending on the type of ground cover (e.g. concrete, sand, gravel).

5.4. Care should be taken to consider secondary factors, especially the meteorological conditions in the region. For example, the ambient temperature will govern the rate of evaporation of a discharged liquid and will control the rate of release of volatile vapours from a pooled liquid, and these processes should be taken into account.

5.5. If a hazardous liquid is volatile (e.g. has a high vapour pressure), such as gasoline, it can give rise to hazardous vapour clouds, whose dispersion as a plume will be consistent with the characteristics of gas cloud dispersion, and this should be considered.

5.6. The mechanisms involved in the dispersion of liquids are such that a release of large quantities of liquid would need to occur for this to directly affect an adjacent nuclear installation. The liquid material will pool and give off toxic or flammable or explosive vapours, and these secondary hazards should be considered as they are likely to pose the most significant hazard to nuclear safety.

5.7. Liquids dispersing underground are typically under high pressure and disperse through fissures and lines of weakness. This dispersion may be strongly directional, and this aspect should be considered.

5.8. Hazardous liquids stored or handled at the nuclear installation will differ from site to site. The safe distances for hazards such as explosion, toxicity and heat flux should be determined and considered in the layout, and appropriate measures for site protection should be taken.

5.9. Where there are multiple nuclear installations on the same site, a possible source of hazardous liquids is likely to be adjacent installations, as these will be nearby and may be sited at the same level as or higher than the host installation and should be considered.

5.10. The dispersion of liquid on bodies of water depends on the characteristics of the liquid (e.g. the density of the liquid compared with the density of water) and the characteristics of the body of water (e.g. sea, river, lake). On standing water bodies, dispersion is slow. In contrast, hazardous liquids in bodies of flowing water may be quickly transported over large distances. The concentration of hazardous liquids at a given distance from the source will depend on the specific situation. In addition to the toxic, corrosive and explosive properties of the liquid, its potential to clog cooling water intakes should also be considered. The effects of prevailing winds on the dispersal of fluids in water should also be considered.

HUMAN INDUCED EXTERNAL EVENTS INVOLVING HAZARDOUS GASES

5.11. Gases, vapours and aerosols from volatile liquids or liquefied gases might, upon release, form a cloud and drift. The drifting cloud might adversely affect the safe operation of the nuclear installation. For example, if hazardous gases permeate the buildings of the nuclear installation, they might pose a hazard to operating personnel or to items important to safety. This could affect the habitability of the control room and other important plant areas and emergency response facilities, and all such potential effects should be considered.

5.12. The most practical method of defence against a hazard of this type is to ensure protection from the potential source by means of distance. Otherwise, design measures such as protective barriers and/or ventilation systems should be provided.

5.13. Clouds of toxic or asphyxiant gases can have severe effects on the personnel of a nuclear installation. Corrosive gases can damage safety systems and might, for example, cause loss of insulation in electrical systems. These matters should be given careful consideration in the evaluation of the hazards.

5.14. Drifting clouds of explosive or flammable gases or vapours can also adversely affect the nuclear installation without entering buildings (e.g. by affecting people and equipment outside the buildings); consequently, suitable protection measures should be taken. Recommendations on protection against explosions and fires are provided in Sections 6 and 7, respectively.

5.15. Local meteorological conditions should be considered conservatively in estimating the danger due to a drifting cloud of hazardous material. In particular, dispersion studies based on probability distributions of wind direction, wind speed and atmospheric stability class should be made. Another consideration is the local topography between the source of HIEEs and the nuclear installation site, especially for dense (heavier than air) gases that will tend to flow downhill in a similar way to liquids.

5.16. For an underground release of hazardous gases or vapours, consideration should be given to escape routes and to seepage effects that might result in high concentrations of hazardous gases in buildings or the formation of hazardous gas clouds within the screening distance value.

5.17. Where there are multiple nuclear installations on the same site, a source of hazardous gases can be the adjacent installations, as these will be nearby and the opportunity for dispersion of the gas plume will be limited, and this should be considered.

HAZARD ASSESSMENT FOR HUMAN INDUCED EXTERNAL EVENTS INVOLVING THE RELEASE OF HAZARDOUS MATERIAL

Identification of sources of HIEEs

5.18. Stationary sources and mobile sources of HIEEs involving the release of hazardous liquids and gases are listed in Table 2 in the Appendix. Recommendations on data collection are provided in Section 4. First, the regions of interest should be located on the basis of generic screening distance values (see Table A–1 in the Annex). Sources of HIEEs within these regions should then be identified. Owing to the uncertainty associated with screening distance values, sources of HIEEs just beyond these regions should also be identified if they contain especially large quantities of hazardous material.

5.19. Data on potential sources of HIEEs should be collected and the distances between the sources of HIEEs and the nuclear installation site should be calculated.

Screening using distance

5.20. Using the source data, simple and conservative calculations should be made and generic screening distance values for the release of hazardous material should be estimated, taking into account that materials originating from liquid or gas sources can travel long distances. Sources that lie farther away from the nuclear installation site than the generic screening distance values can be screened out. Meteorological and topographical considerations should be taken into account.

Screening using probability

5.21. If a hazard cannot be screened out using distance, generic event data (i.e. based on the total occurrence frequency of an event category) can be used. Pragmatic and conservative judgement should be applied to determine the probability of potential events involving the release of a hazardous fluid. If the total probability of occurrence is less than the screening probability level, it can be screened out. The screening of each source that could lead to the leakage of a hazardous fluid at the nuclear installation site should be completed, and all the screened-in sources should be listed.

Detailed evaluation

5.22. Hazard analysis of screened-in sources should be performed to check the interaction with the nuclear installation. If there is an interaction, hazard characterization is required to be performed (see Requirement 7 and para. 4.19 of SSR-1 [1]).

5.23. In broad terms, the evaluation process should consider the release of a hazardous liquid at a specified location in terms of leak rate and possibly other factors if storage was not at ambient atmospheric conditions. The evolution of the release is driven by local topography for overland releases and by the local marine or watercourse conditions for releases into the hydrosphere. These aspects should be modelled explicitly, or else conservative assumptions should be made. Liquids released into the hydrosphere and gases emanating from liquids are extremely important and should be considered.

5.24. Vapour clouds released after an event can travel to the nuclear installation site and might cause damage to items important to safety or might affect the habitability of the control room. Different chemicals have different hazardous effects relating to explosion, thermal radiation and toxicity. In the evaluation, the worst case meteorological conditions should be assumed as inputs to the model within bounding conditions of temperature, atmospheric stability class and wind speed for each chemical modelled and each hazard condition until the maximum potential effect is confirmed.

5.25. The nearest point to the nuclear installation where hazardous liquids might collect in pools should be determined, with account taken of the topography of the land and the layout of the installation. Similarly, a gas release should be modelled by assuming a maximum credible inventory that occurs at the point of closest approach to the nuclear installation site (or the most unfavourable release point, if this is different). Mobile sources, such as barges and ships carrying large amounts of hazardous liquids or gases within the generic screening distance, should be assumed to become stranded at the point of approach to the nuclear installation at which the most unfavourable effects would result.

5.26. For evaluating the generation of hazardous gases, vapours and aerosols and the interaction with the nuclear installation, a distinction should be made between subcooled liquefied gases, gases liquefied by pressure and non-condensable compressed gases.

5.27. Usually, the release of a subcooled liquefied gas occurs as a steady leak over a considerable period (at a given leak rate), but the possibility of an effectively instantaneous release (i.e. a total sudden release) should also be considered, depending on the following conditions associated with the release:

(a)   The type of storage container and its associated piping;
(b)   The maximum size of the opening from which the material might leak;
(c)   The maximum amount of material that might be involved;
(d)   The relevant circumstances and mode of failure of the container.

5.28. The starting point is the evaluation of a range of leak rates and related failure probabilities or of the total amount of gas released (equivalent to the maximum credible release) and related failure probability. If a large amount of subcooled liquefied gas is released, much of it might remain in the liquid phase for a long time. It should be treated as a liquid throughout this period, although a small fraction will vaporize almost instantaneously. The characteristics of the pool formed by the liquid, such as its location, surface area and evaporation rate, should be evaluated, with account taken of the permeability and thermal conductivity of the soil (if the spillage occurs on soil). If the source site has arrangements for containing any spills or releases, these should be taken into account. However, giving credit to such arrangements should be justified.

5.29. To evaluate the maximum concentration of neutral buoyant gases at the site, the models presented in IAEA Safety Standards Series No. NS-G-3.2, Dispersion of Radioactive Material in Air and Water and Consideration of Population Distribution in Site Evaluation for Nuclear Power Plants [5], can be used. However, specific models should be used for heavy gases.

5.30. The formation of a large cloud is more likely for gases liquefied by pressure and for non-condensable compressed gases than it is for subcooled liquefied gases. The detailed evaluation of gases liquefied by pressure and of non-condensable compressed gases is easier because the source is more easily defined and, in some cases, dispersion of the plume is governed by simpler phenomena. As with subcooled liquefied gases, the release of gases liquefied by pressure and of non-condensable compressed gases should be characterized by a leak rate or by a sudden total release, and a similar evaluation should be carried out. The assumptions to be used will depend on the type of storage tank, the process vessels, their associated piping, pipelines with associated flow rate and operating pressure, and the associated failure probability.

5.31. In making assumptions about the amount of material available to be released in the event of an accident, account should be taken of the time interval before action is taken to stop the leak. For example, pipeline valves may close automatically, thus isolating the ruptured section quickly.

5.32. With buried pipelines, the soil cover is usually insufficient to prevent the escape of released gases. Seepage might occur or gas might escape through fractures or discontinuities. In all cases, when the characteristics of the gaseous release to the atmosphere have been established, a model should be selected to determine the dispersion of the gas toward the nuclear installation site. The dispersion of the plume is primarily governed by the meteorological conditions at the time of release. Given the large degree of uncertainty associated with meteorological and other factors involved in plume modelling, consideration should be given, at least initially, to using a simplified dispersion model with conservative assumptions.

Hazard parameters

5.33. The following are examples of hazard parameters that should be considered in relation to the release of hazardous material (see Table 2 in the Appendix):

(a) Nature of material:
   — Physical properties:
     • Density, temperature and pressure, as contained;
     • Density, temperature (including freezing and boiling temperatures), partial vapour pressures under ambient conditions;
     • Flow characteristics under ambient conditions.
   — Chemical properties:
     • Composition;
     • Reactivity with environmental and atmospheric materials.
(b) Radiochemistry.
(c) Flashpoint or ignition temperature.
(d) Maximum credible release, or frequency versus quantity release relationship. This involves gathering data and parameters in relation to the storage or process, such as dimension, horizontal or vertical storage, maximum pressure rupture, height and shape of the release. In the case of a chemical

reaction leading to a release, the release rate due to the chemical reaction as well as the location of the source release (i.e. size and height of the stack) should be known.

(e) Meteorological and topographical characteristics of the region.
(f) Bathymetric and tidal characteristics of the coastal region.
(g) Watercourse and flooding characteristics of the fluvial region.
(h) For underground sources, geological seepage routes and opportunities for liquid concentration.
(i) Existing protective and mitigatory measures at the source location.
(j) Type of the soil and subsoil (e.g. nature, roughness, permeability).

Load characterization parameters

5.34. The following are examples of load characterization parameters that should be considered (see Table 4 (5) and (6) in the Appendix):

(a) Asphyxiant or toxic materials:
— Concentration and quantity as a function of time;
— Volatility in ambient conditions;
— Toxicity and asphyxiant limits.
(b) Corrosive or radioactive liquids:
— Concentration and quantity as a function of time;
— Corrosiveness and radioactive content.
(c) Location of material (e.g. over or in the sea, overground or underground).

6. HUMAN INDUCED EXTERNAL EVENTS INVOLVING EXPLOSIONS

6.1. The word 'explosion' is used in this Safety Guide broadly to mean any exothermic chemical reaction between solids, liquids, vapours or gases that could cause a substantial increase in pressure, owing to impulse loads, drag loads, fire or heat, and/or a rapid release of a liquid or gas from a pressurized container. The explosive potential of a given mass of chemical material is often quoted in terms of an equivalent mass of trinitrotoluene (TNT). This facilitates comparison of the explosive potential of different materials, and many empirical formulas for predicting the effects of explosives are derived on the basis of TNT equivalence [18]. These should be used with care as described in para. 6.18.

6.2. Explosions are highly energetic and often destructive events, and they can occur for many reasons. Once an explosion has occurred, its effects are propagated into the surrounding environment by means of an expanding pressure wave. There are two types to consider, as follows:

(a) Deflagrations, which generate moderate pressure waves, heat and fire;
(b) Detonations, which generate high near field pressure waves and associated drag loading, usually without significant thermal effects.

These pressure waves, also known as blast waves, propagate approximately as spherical waves expanding away from the source location and should be considered. However, they are influenced by the ground and other confining surfaces. The specific energy in a spherical wave front attenuates in accordance with the inverse square law based on distance from the source if no further energy is being added (e.g. by continued burning) to the wave. However, constrained blast waves may attenuate much more slowly[9]. More details are provided in Ref. [18].

6.3. Explosions at an industrial site usually occur owing to overpressurization of contained liquids and/or gases, or to deflagrations of liquid pool fires, leaks from or failure of storage tanks and pipelines, runaway chemical reactions or accidents with explosives. In addition, dust explosions can also occur where any dispersed powdered combustible material is present in sufficient concentrations. In underground operations, outbursts of natural gases such as methane can create explosions. Explosions due to any cause should be considered.

6.4. Explosions normally arise from hazardous (often flammable) materials and the way they are contained or handled. The release of hazardous material is addressed in Section 5. The ways in which explosion hazards affect structures, systems and components and personnel at a nuclear installation are covered in detail in other Safety Guides (e.g. SSG-68 [7]), but the propagation phenomena from the source to the nuclear installation site are addressed in this section.

---

[9] The attenuation referred to is geometric attenuation, as this is normally the most significant effect. For comparison purposes, cylindrical waves geometrically attenuate as the inverse of distance from the source, and one-dimensional waves do not attenuate at all. Blast waves will also suffer viscous attenuation with time of travel, but this phenomenon is relatively slow acting. Attenuation refers to the energy of the wave front. Since energy is related to the square of particle velocity and strain, these parameters attenuate as the square root of energy.

6.5. An overpressurization event is an event arising from an overpressurized container of a liquid or gas that can result in an explosive release of the liquid or gas if the container fails. When such a release is also associated with heating, or the released material ignites, the result can be an extremely energetic form of release known as a boiling liquid expanding vapour explosion. This can occur in all types of contained materials, but generally occurs when tanks containing pressurized liquid petroleum gas, liquid natural gas or propane fail catastrophically. If such tanks are accidentally heated, as might be the case if they are immersed in an external fire, the pressure in the tank rises until it eventually bursts. The mechanical overpressure effects of the burst itself may be sufficient to cause a boiling liquid expanding vapour explosion, but if liquid natural gas vapour ignites, this adds substantially to the energy of the explosion and can lead to an extremely destructive event, characterized by a detonation blast wave, and should be considered. Damage due to projectiles created by a boiling liquid expanding vapour explosion should also be considered.

6.6. In the case of a hydrocarbon liquid pool or similar scenarios, the hydrocarbon can escape the containment, form a vapour cloud and ignite (known as a vapour cloud explosion). In flammable atmospheres, the explosion pressure wave is characterized by a flame front. The speed of propagation of the flame front depends on the availability and rate of burning of the fuel source (e.g. petroleum vapour). These events generally produce deflagration pressure waves and should be considered.

6.7. Dust explosions are especially dangerous and can easily lead to detonations because of the rapid rate of combustion of fine particles. The rate of combustion is related to the surface area of fuel in contact with air, so a large number of fine particles (or vapour droplets from such particles) burns more effectively than a small number of larger ones. The presence of obstacles that are often found in powder stores (e.g. grain stores) can cause intense mixing as the blast wave propagates, leading to more rapid burning and hence a more intense blast wave, often with very dramatic effects, and should be considered. A hybrid explosion can be difficult to predict because the data are normally only available for separated materials (e.g. an ignited cloud containing a mixture of gas and dust). Such an explosion can cause more intensive effects depending on changes in the mixture (e.g. lower limit of explosion and maximum pressure). Particular attention should be given to such potential hybrid explosions.

6.8. Blast waves cause a sudden increase in pressure on one side of a structure with insufficient time for pressure on the other side to equalize through the action of normal ventilation processes. This results in large pressure forces across the surface of the affected structure, and hence large stresses should be considered.

6.9. An explosion can produce pressure waves (normally the dominant hazard), projectiles, heat, smoke, dust and ground shaking. A vapour cloud explosion is also possible if relevant conditions are met, and this should also be considered.

6.10. Explosions are very likely to create secondary hazards. For example, structural damage close to the event can generate projectiles, destroy critical infrastructure and initiate fire. Secondary hazards associated with explosions should be considered.

6.11. A significant factor affecting the propagation of blast waves is the presence of obstacles between the source of the HIEEs and the nuclear installation site, and inside the vapour cloud; local topography and the layout of the site may also play a role, and both effects should be considered.

6.12. The interactions between units collocated at a site containing multiple nuclear installations should be carefully considered for their contribution to HIEE explosion hazards.

6.13. Particular attention should be paid to potential hazards associated with large explosive loads such as those transported by freight trains or in ships.

6.14. Unless there is adequate justification, a conservative assumption should be made that the maximum amount of explosive material usually stored at the source of HIEEs will explode, and an analysis should then be made of the effects of the resulting hazards (e.g. incidence of pressure waves, ground shock, projectiles) on the nuclear installation. The secondary effects of fires resulting from explosions should also be considered, as described in Section 7.

6.15. The probability with which explosions might occur should be calculated on the basis of operating experience or be derived from national or worldwide data. More information on explosion hazards can be found in Ref. [18].

HAZARD ASSESSMENT FOR HUMAN INDUCED EXTERNAL EVENTS INVOLVING EXPLOSIONS

Identification of sources of HIEEs

6.16. Sources of HIEEs involving explosions are listed in Table 2 in the Appendix. Recommendations on data collection are provided in Section 4. First, the regions of interest should be located on the basis of generic screening distance values (see Table A–1 in the Annex). Sources of HIEEs within these regions should then be identified. Owing to the uncertainty associated with screening distance values, sources of HIEEs just beyond these regions should also be identified if they are especially hazardous.

6.17. Data on potential sources of HIEEs should be collected and the distances between the sources of HIEEs and the nuclear installation site should be calculated.

Screening using distance

6.18. Using source data, generic screening distance values for overpressure (the dominant hazard) should be estimated by means of a simplified conservative approach based on the engineering relationship between the TNT equivalent mass and the distance. This is applicable for high explosives with the potential for mass casualties. For hydrocarbon–air vapour cloud explosions, other appropriate methodologies should be used. Sources of explosion can be screened out if they are farther away from the nuclear installation site. Meteorology, topography and existing protective measures at the source should be taken into account.

Screening using probability

6.19. If a hazard cannot be screened out by distance, generic event data (i.e. based on the total occurrence frequency of an event category) can be used. Pragmatic and conservative judgement should be applied to determine the probability of a potential event that could create an explosion. If the total probability of occurrence is less than the screening probability level, it can be screened out. Appropriate methods for calculating the probability of an explosion should be used. If there are not enough statistical data available for the region to perform an adequate analysis, reference should be made to global statistics, to pertinent data from similar regions and/or to expert judgement including site visits. The screening of each source that could create a pressure wave at the nuclear installation site should be performed and the screened-in sources should be listed.

Detailed evaluation

6.20. Hazard analysis of screened-in sources should be performed to check the interaction with the nuclear installation. If there is an interaction, hazard characterization is required to be performed (see Requirement 7 and para. 4.19 of SSR-1 [1]).

6.21. In this step, the list of screened-in hazards should be refined by a more detailed assessment of the range of potential events for their applicability to the specific nuclear installation. Typical screening parameters that should be applied in this step are design robustness, distance, magnitude, probability and zones of influence.

6.22. The pressure waves, drag level and local thermal effects at the nuclear installation will differ depending on the nature and amount of explosive material, the configuration of the explosive material, meteorological conditions, the layout of the nuclear installation and the topography. Certain assumptions are usually made to develop the design basis for explosions, with data on the amounts and properties of the chemicals involved taken into account. TNT equivalents are commonly used as a first approach to estimate safe distances for given amounts of explosive chemicals and for a given pressure resistance of the structures concerned. This is applicable for high explosives with potential for mass casualties. For hydrocarbon–air vapour cloud explosions, other appropriate methodologies should be used. For certain explosive chemicals, the pressure–distance relationship has been determined experimentally and should be used directly.

6.23. Projectiles that might be generated by an explosion should be identified by using operating experience data and engineering judgement on the source of these projectiles. In particular, the properties of the explosive material concerned and the characteristics of the facility in which the explosion is assumed to occur should be considered.

6.24. Consideration should also be given to possible ground motion and to other secondary effects such as the outbreak of fire, the release or production of toxic gases and the generation of dust.

Hazard parameters

6.25. The following are examples of hazard parameters that should be considered (see Table 2 in the Appendix):

(a) Nature of the explosive material:
  — Physical properties;
  — Chemical properties;
  — Radiochemistry;
  — Flashpoint or ignition temperature.
(b) Maximum credible pressure and thermal release, or the relationship between the frequency of explosion and the severity.
(c) Meteorological and topographical characteristics of the region.
(d) Existing protective and/or mitigative measures at the source location.
(e) Parameters for the determination of the release rate of the flammable source (e.g. evaporation rate in the case of a flammable pool of hydrocarbon, release rate for a flammable gas release).

Load characterization parameters

6.26. The following are examples of load characterization parameters that should be considered (see Table 4 (1), (2), (3) and (4) in the Appendix):

(a) Overpressure as a function of time.
(b) Hard and soft missiles.
(c) Heat: maximum temperature flux and duration.
(d) Smoke and dust:
  — Composition;
  — Concentration and quantity as a function of time.
(e) Ground shaking: frequency response spectrum for vibrational motion.

7. HUMAN INDUCED EXTERNAL EVENTS INVOLVING FIRE

7.1. There are several possible sources of external fire that could threaten a nuclear installation, including fires starting in adjacent units or installations on the same site. Fires from aircraft crashes are addressed in detail in Ref. [18].

7.2. A survey should be made at and around the site to identify potential sources of fire, such as forests, vegetation and peat; storage areas for flammable materials (especially hydrocarbon storage tanks), wood and plastics; factories that produce or store such materials and their transport routes; pipelines and chemical plants; and accidents on major highways. Fires can be accompanied by other hazards such as explosion and release of hazardous material because of their ability to cause the failure of containment structures such as tanks. Fire is often also a secondary or consequential hazard following such events.

7.3. Depending on the nature and properties of the flammable material (e.g. volatility, physical state, storage conditions, release type), different fire phenomena can be observed, such as pool fire, jet fire, fireball and vapour cloud explosion. These events could occur simultaneously or sequentially, and this should be taken into consideration.

7.4. Fire can spread horizontally in different ways: by radiation heating from the thermal flux associated with the fire, via flammable material situated between the fire source and the site or installation, or by sparks. Significant passive protection can be provided by the presence of fire breaks and/or by ensuring that areas immediately external to the site or installation are free of flammable material. In the case of external fires, alternative fire spread paths should also be identified, such as airborne dispersion of firebrands (embers) or transportation of liquid fuel in the sewer system.

7.5. The heat flux in quiescent conditions will obey the inverse square law of energy attenuation; however, some fire related hazards such as smoke may propagate directionally owing to the prevailing wind direction and attenuate slowly in this direction. The fire itself will spread preferentially in the downwind direction, especially if there is a supply of flammable material along the route such as dry vegetation. All these factors should be considered.

7.6. Nuclear installations may have a substantial ability to withstand thermal heating before the safety of the installation is affected; however, smoke could quickly affect safety if, for example, it prevents operating personnel from performing an important safety function or blocks an air filter. Sites containing multiple nuclear installations should be considered carefully for fire hazards due to HIEEs. Thermal heating from an external fire can also create a secondary hazard, for example structural damage creating a leak that leads to a release of hazardous material. Secondary hazards associated with thermal heating should be considered.

7.7. The protective measures against fire hazards taken at the nuclear installation and at the source of the fire should be considered in evaluating the effects of external fires on the nuclear installation. However, before giving credit to these measures in the hazard evaluation, sufficient justification should be provided.

HAZARD ASSESSMENT FOR HUMAN INDUCED EXTERNAL EVENTS INVOLVING FIRE

Identification of sources of HIEEs

7.8. Sources of HIEEs involving fire are listed in Table 2 in the Appendix. Recommendations on data collection are provided in Section 4. First, the regions of interest should be located on the basis of generic screening distance values (see Table A–1 in the Annex). Sources of HIEEs within these regions should then be identified. Owing to the uncertainty associated with screening distance values, sources of HIEEs just beyond these regions should also be identified if they are especially hazardous.

7.9. Data on potential sources should be collected and the distances between the sources of HIEEs and the nuclear installation should be calculated. Sources to be considered include forests, peat, vegetation, storage areas for low volatility flammable materials (especially hydrocarbon storage tanks), industrial facilities that process flammable materials and associated transport routes.

Screening using distance

7.10. Using the source data, screening distance values for heat flux (the dominant hazard) can be estimated by means of a simplified conservative approach. Sources of fire that lie farther away from the nuclear installation site can be screened out. Meteorology, topography and existing protective measures at the source and nuclear installation should be taken into account.

Screening using probability

7.11. If a fire hazard cannot be screened out by distance, generic event data (i.e. based on the total occurrence frequency of an event category) can be used. Pragmatic and conservative judgement should be applied to determine the probability of potential events that could initiate a fire. If the total probability of occurrence is less than the screening probability level, it can be screened out. The screening of each source that could lead to fire at the nuclear installation site should be completed, and all of the screened-in sources should be listed.

7.12. If the potential fire hazard from screened-in sources of HIEEs is likely to be less than that from similar materials stored on the nuclear installation site and against which protection has already been provided, then these sources can be screened out. If several sources are screened out on the same basis, it may be necessary to reflect the frequency contribution arising from the sum of all such sources by nominating a bounding source and screening using this. Each event that could initiate a fire and affect the nuclear installation site should be screened. The screened-in sources should be listed.

Detailed evaluation

7.13. Hazard analysis of screened-in sources should be performed to check the interaction with the nuclear installation. If there is an interaction, the hazard is required to be evaluated (see Requirement 7 and para. 4.19 of SSR-1 [1]). In this step, the list of screened-in hazards should be refined by a more detailed assessment of the range of potential events for their applicability to the specific nuclear installation. Typical screening parameters that should be applied in this step are design robustness, distance, magnitude, probability and zones of influence.

7.14. The hazard evaluation should consider the location of the source of fire and assume a type of fire and/or flammable material and ignition mechanism. The probability of fires can be obtained from operating experience or be derived from general national or worldwide data.

7.15. To avoid fire effects from forests and/or bushes, it should be ensured that a zone around the nuclear installation site is devoid of any vegetation. A fire safety programme at the site should be implemented to avoid fires from other sources that could affect the safety of the nuclear installation.

7.16. The thermal exposure of external structures, systems and components at the nuclear installation should be quantified in terms of the radiative and convective heat flux incident on the target surface and the duration of the exposure. Methods to assess external fireballs and pool fires from a sudden release and ignition of combustible liquid or gas are provided in Ref. [18]. Smoke is another important hazard that should be evaluated, including the potential for it to travel longer distances.

Hazard parameters

7.17. The following are examples of hazard parameters that should be considered (see Table 2 in the Appendix):

(a)   Nature of the flammable material and its source:
         — Flashpoint, flammability concentrations in air or other ignition criteria;

— Maximum credible material release or thermal release, or the relationship between fire frequency and severity;
— Thermal load as a function of time.

(b) Meteorological and topographical characteristics of the region.
(c) Existing protective measures at the source location (e.g. fire breaks).

Load characterization parameters

7.18. The following are examples of load characterization parameters that should be considered (see Table 4 (3), (4) and (5) in the Appendix):

(a) Heat: maximum heat flux and duration.
(b) Smoke and dust:
— Composition;
— Concentration and quantity as a function of time.
(c) Asphyxiant and toxic material:
— Concentration and quantity as a function of time;
— Volatility in ambient conditions;
— Toxicity and asphyxiant limits.

8. HUMAN INDUCED EXTERNAL EVENTS INVOLVING AN AIRCRAFT CRASH

8.1. Methods currently in use for considering an aircraft crash as an HIEE may contain differences in terms of detail; however, they all contain the same basic elements that should be considered, as follows:

(a) Categorization of aircraft by type, mass, velocity and size.
(b) Categorization of airspace by the type of flying rules or restrictions that apply (e.g. commercial airways, airspace around airports, restricted airspace).
(c) Frequency analysis to determine the crashes per year per square kilometre at the location of the nuclear installation site for each aircraft category.
(d) Frequency analysis to determine the aircraft crash into a nuclear installation that could lead to a radioactive release. This includes calculating the area of the nuclear installation site that is variously referred to as the target area, zone of influence or damage footprint. In some States, the crash of a large passenger airplane is postulated independently of the actual crash probability.

8.2. Aircraft should be considered to be a mixture of hard and soft missiles whose impact on reinforced concrete structures typically results in damage modes such as perforation, penetration, scabbing, spalling, local punching, bending failure and vibrations.

8.3. In some nuclear installations, specific protection is provided against malicious aircraft crashes; such protection measures are generally sufficient to envelop the risk from accidental aircraft crash hazard, such that it can be screened out. Nevertheless, it should be carefully checked whether the assumed scenarios for malicious aircraft crashes fully cover potential accidental scenarios and whether the protection measures are suitable for accidental aircraft crashes. Malicious aircraft crashes are not considered in this Safety Guide; however, some of the methods recommended in this Safety Guide may also be applicable to the evaluation of hazards from a malicious aircraft crash when such a scenario cannot be screened out.

8.4. An aircraft crash is potentially one of the most significant of all HIEEs, and a large amount of research has been conducted into the methods for crash probability analysis and the effects of impact events onto heavy concrete targets. This research and experience should be considered in the aircraft crash hazard evaluation.

8.5. It is important to consider all the potential effects of an aircraft crash on the nuclear installation if such an event is not screened out, as follows:

(a) Direct effects:
         — Impact damage to structures, including perforation and penetration;
         — Vibration effects;
         — Global stability.
(b) Secondary effects:
         — Secondary missiles ejected from the impact site and scattering widely;
         — Rapid spread of flammable liquid from the point of impact;
         — Entry of combustion products into ventilation or air supply systems;
         — Fire and explosion generating heat and blast effects and tertiary missiles;
         — Release of hazardous material carried as cargo.

8.6. The main component in the loading function resulting from a collision of the deformable fuselage can be predicted assuming a soft missile impact. Aircraft engines and landing gear can be classified as semi-hard or hard missiles and should be considered.

8.7. Fire from fuel spillage can result in fireball or pool fire or both and should be considered. Combustible cabin materials, payloads and carbon fibre based structural materials will also be involved in fire and should be counted as fire loads. Details are provided in Ref. [18].

8.8. At sites on which multiple nuclear installations are located, there may be multiple items important to safety serving different units. An impact on structures associated with an adjacent unit might not directly impact the unit under consideration, but secondary hazards such as missiles, fire and explosion should be considered.

HAZARD ASSESSMENT FOR HUMAN INDUCED EXTERNAL EVENTS INVOLVING AN AIRCRAFT CRASH

8.9. Air traffic encounters several different operational environments that critically affect the probability of crash events. The following types of aircraft crash should be considered:

— Type 1: Aircraft crash arising from general air traffic, sometimes called the background crash rate.
— Type 2: Aircraft crash arising from take-off and landing manoeuvres at a local airport.
— Type 3: Aircraft crash arising from air traffic in the main civil traffic corridors and military flight zones.

Type 1 aircraft crash

Identification of sources of HIEEs

8.10. Information on aircraft crashes in the State should be collected from the civil and military aviation authorities and/or other national authorities working in the aviation industry. This information should include details of crashes of all types of aircraft flying in the State.

Screening using distance

8.11. Screening using distance is not applicable for this type of event.

Screening using probability

8.12. Aircraft crash data covering a regional circular area (e.g. typically 100–200 km in radius) for each type of civil aircraft crash and military aircraft crash should be considered. The probability of Type 1 crashes should be evaluated, in particular in densely populated regions with more than one civil airport and with large numbers of flights. Appropriate zoning of the area considered should be carried out to ensure that any averaging is sufficiently conservative.

8.13. The probability of occurrence of all types of aircraft crash should be evaluated by considering the site as a tract or circular area, by dividing the site area by the regional area and multiplying by the average number of aircraft crashes per year for different types of aircraft (e.g. typically $0.1–1$ km^2). Those types of aircraft for which the probability of crashing is less than the screening probability level can be screened out. Other types of aircraft should be retained for detailed evaluation.

Type 2 aircraft crash

Identification of sources of HIEEs

8.14. Sources of HIEEs involving a Type 2 aircraft crash are listed in Table 2 in the Appendix, and an associated generic screening distance value is shown in Table A–1 in the Annex. Recommendations on data collection are provided in Section 4. The probability of an aircraft crash is usually higher in the vicinity of airports, both civil and military. The identification of sources of HIEEs should be performed separately for both types. Most aircraft crashes tend to occur within approximately semi-circular areas centred at the ends of the runways (e.g. typically 8 km in radius, as shown in Table A–1 in the Annex).

Screening using distance

8.15. If regional or national values have been specifically established, they can be used. Otherwise, generic screening values should be used.

Screening using probability

8.16. If a hazard cannot be screened out by distance, the total probability of occurrence for each one of the aircraft categories (e.g. general aviation, medium and large commercial aircraft, military aircraft) should be determined and compared with the screening probability level. Those aircraft crashes for which the probability of occurrence is less than the screening probability level can be screened out. Otherwise, the crashes should be subject to detailed evaluation. Certain crash characteristics, such as a lower impact angle and lower impact speed, can be considered during that stage.

Type 3 aircraft crash

Identification of sources of HIEEs

8.17. Sources of HIEEs involving a Type 3 aircraft crash are listed in Table 2 in the Appendix, and an associated generic screening distance value is shown in Table A–1 in the Annex. Recommendations on data collection are provided in Section 4. If airways or airport approaches pass over the site (or within 4 km of the site (see Table A–1 in the Annex)), the potential hazards arising from aircraft crashes due to air traffic in the main civil traffic corridors and military flight zones should be considered. The screening should be based on the distance from the edge of the flight zone, air traffic corridor or approach, as appropriate.

Screening using distance

8.18. If regional or national screening distance values have been established, they can be used. Otherwise, generic screening distance values should be used.

Screening using probability

8.19. If a hazard cannot be screened out by distance, the total probability of occurrence of each of the aircraft categories (e.g. general aviation, medium and large commercial aircraft, military aircraft) should be determined and compared with the screening probability level. Those aircraft crashes for which the probability of occurrence is less than the screening probability level can be screened out. Otherwise, the crashes should be retained for detailed evaluation.

Detailed evaluation for all types of event

8.20. Hazard analysis should be performed for the screened-in sources and hazards are required to be characterized (see Requirement 7 and para. 4.19 of SSR-1 [1]).

8.21. In this step, the list of screened-in hazards should be refined by more detailed assessment of the range of potential events for their applicability to the specific nuclear installation. This assessment should be based on the specific characteristics of the site and the nuclear installation. Typical parameters that should be applied are design robustness, distance, magnitude, probability and zones of influence. An additional consideration is the type and number of collocated installations on the site that can have positive or negative effects on prevention, detection, control of consequences (in normal and severe conditions) and emergency response. Details are provided in Ref. [18].

8.22. An approach similar to the zone of influence should be used. The concept of defining areas of consequence for each hypothetical impact location should be employed. The areas of consequence are denoted as damage footprints. Damage footprints are defined for impact, shock and fire loading conditions.

8.23. The evaluation of hazards from an aircraft crash should consider the buildings containing nuclear material and the buildings housing structures, systems and components important to safety (e.g. equipment for heat removal), as follows:

(a)   Impact locations to be considered should be defined on the basis of the aircraft parameters (e.g. type of aircraft, nature of flight, angle of impact), shielding by topography, nuclear installation buildings, transmission lines and other considerations.
(b)   Conservative assumptions about the angle of aircraft impact (e.g. perpendicular to the centreline of the containment building, perpendicular to the spent fuel storage building) should be made.
(c)   Local response, global response and vibration loading conditions should be considered.

8.24. All buildings of the nuclear installation containing the structures, systems and components necessary to protect against the hazards associated with an aircraft crash should be identified for further screening or for evaluation. For example, systems and support systems needed for safe shutdown of a reactor or continued cooling of a spent fuel pool should be identified. The exterior faces of the buildings should be evaluated to screen out the need for further evaluation or to determine impact locations, as follows:

(a)   Faces or partial faces of buildings could be screened out from further consideration owing to shielding by adjacent structures, intervening structures or other site features.
(b)   Faces of buildings that are partially screened out should be subdivided into those portions for which aircraft impact is possible and those for which such impact is not possible.
(c)   The impact on multiple buildings during the event should be considered, in order to identify multiple buildings vulnerable to a single aircraft crash.

Damage footprints for each building and each impact location on the buildings should be developed for evaluation.

8.25. After evaluation, loading functions for the screened-in HIEEs should be defined for the engineering evaluation. The load characterization is the link between the events and the definition of the loading environment for evaluation. The resulting matrix of loading conditions produced by the events should be applied to the entire nuclear installation or to portions of it (see table 4, Scenario No. 1, in Ref. [19]). Tables 5–7 of Ref. [19] identify the following parameters for engineering evaluation: impact, heat, fire and vibration. Reference [18] describes the complete evaluation methodology for structural impact, induced vibrations, thermal effects from fire, local and global effects and acceptance criteria.

Hazard parameters

8.26. The following are examples of hazard parameters that should be considered in relation to aircraft crashes (see Table 2 in the Appendix):

(a)   Types of aircraft and characteristics, nature of flight and crash rate.
(b)   Aircraft movements and flight frequencies from or in the following:
      — Airports;
      — Airways;
      — Controlled airspace around commercial and military airports;
      — Restricted and other forms of special airspace;
      — Location of aircraft sources, runway directions and other related data, and direction of approach to the nuclear installation site;
      — Airfield plates[10] for take-off, landing and manoeuvring.
(c)   Parameters derived from regional or national aircraft crash data:
      — Probability distributions for direction of approach and angle of descent for each aircraft type;
      — Skid and footprint distances and rate of energy and momentum attenuation with distance for each aircraft type.

---

[10] Airfield plates (paper based and now digital) provide all the navigational information needed by a pilot manoeuvring around a major airport. They are prepared by national authorities and specific to the airport, runway, runway direction and navigational procedure being used. They are publicly available for all international airports and many local ones.

Load characterization parameters

8.27. The following are examples of load characterization parameters that should be considered (see Table 4 (1), (2), (3), (4) and (6) in the Appendix and Ref. [18]):

(a) Impact energy at the nuclear installation:
— Mass;
— Velocity.
(b) Impact parameters:
— Components of aircraft classified as hard missiles and as soft missiles;
— Size and cross-section area of plane of impact.
(c) Parameters derived from the hazard analysis:
— Probability distributions for direction of approach and angle of descent onto the nuclear installation site for each aircraft type;
— Skid and footprint distances and rate of energy and momentum attenuation with distance for each aircraft type;
— Data needed for analysis of secondary hazards;
— Fuel load for each aircraft type and stage of flight;
— Hazardous cargo, materials and volumes.

9. HUMAN INDUCED EXTERNAL EVENTS INVOLVING TRANSPORT, EXCLUDING AIRCRAFT CRASHES

9.1. Mobile sources, excluding air traffic, of HIEEs include the following (see Tables 2 and 3 in the Appendix):

(a) Road transport: trucks carrying hazardous material.
(b) Rail transport: trains carrying hazardous material.
(c) Marine transport:
— Ships carrying hazardous material (cargo);
— Ships that possess significant kinetic energy.
(d) River transport:
— Barges carrying hazardous material (cargo);
— Barges that possess significant kinetic energy.
(e) Pipelines: pipelines conveying hazardous material.

9.2. This section considers some general features of road, rail, sea and waterway transport events before dealing collectively with all sources that present a direct impact hazard to the structures, systems and components of a nuclear installation, and with those that can lead to a release of hazardous material.

9.3. Vessels have the potential to interact with coastal and offshore structures belonging to a nuclear installation site. Damage to nuclear reactor cooling water intakes and outfalls is a potential concern, as is potential damage to docks and jetties that are used for loading and unloading nuclear material onto vessels. The potential for vessels to interact with coastal and offshore structures of a nuclear installation should be considered.

9.4. Road and rail vehicles and marine and river vessels routinely transport hazardous material, and the release of hazardous material is always a potential risk to nearby nuclear installations and should be considered. All hazards should be addressed in accordance with the recommendations provided in the previous sections by taking the closest distances from the nuclear installations. Similarly, pipelines routinely convey hazardous liquids or gases and should also be considered.

MARINE AND RIVER VESSELS THAT POSSESS SIGNIFICANT KINETIC ENERGY

9.5. The effects on a nuclear installation of marine and river vessels that possess significant kinetic energy will depend on the nature of any shoreline and offshore structures, their layout and whether there is any natural or human-made protection. The most significant event is a collision between a massive vessel and a shoreline (e.g. dock, loading facility) or submerged safety structure (e.g. cooling water intake), where substantial structural damage is possible. Such events can be regarded as soft missile impacts, where significant deformation of both the vessel and the coastal structure is likely and should be considered.

9.6. The primary hazard is impact, but secondary effects of oil spill, fire, explosion and release of gases are possible and should be considered in accordance with the recommendations provided in the previous sections. Other cargo that is not formally classified as hazardous material, such as thick liquids, pastes, absorbent bulky freight (e.g. wood pellets) and sticky chemicals, should also be considered, as it might jeopardize the water intake.

9.7. Large commercial ships can drift by tide and river currents. The local bathymetry around the nuclear installation should be considered, and tide and river flow conditions should be selected to identify the most onerous conditions of vessel reach and speed relative to the nuclear installation structures.

HAZARD ASSESSMENT FOR HUMAN INDUCED EXTERNAL EVENTS INVOLVING MARINE AND RIVER VESSELS THAT POSSESS SIGNIFICANT KINETIC ENERGY

Identification of sources of HIEEs

9.8. Sources of HIEEs involving marine and river vessels are listed in Table 2 in the Appendix. Recommendations on data collection are provided in Section 4. Data on potential sources of HIEEs should be collected and the distances between these sources and the nuclear installation site should be calculated. Data collection should include information on ships and barges entering the loading and unloading area of the site, commercial vessels moving in designated shipping lanes and maintenance vessels used for dredging. Information should be collected from local marine and river authorities on the location of shipping lanes, the local bathymetry, tide and river flows throughout the year, and on the frequency and nature of vessel movements.

Screening using distance

9.9. Using the collected data on the sources of HIEEs and on the protective measures at the nuclear installation site, it should be determined whether any vessel could impact an intake structure. Each vessel needs a certain water depth to move and reach the coast. Local bathymetry, predominant tide and wind direction are important considerations, but worst met conditions should also be considered. If a vessel cannot impact any structures important to safety, the hazard can be screened out.

Screening using probability

9.10. If a hazard cannot be screened out by distance, generic event data (i.e. based on the total occurrence frequency of an event category) can be used. Pragmatic and conservative judgement should be applied to determine the probability of an event that can initiate an impact. If the total probability of occurrence is less than the screening probability level, it can be screened out. The probability of a collision with a commercial vessel with the water intake structure could be very low if protective embankments are constructed with an opening for the cooling water. Vessels entering the intake channel can impact the intake structure as a result of human error if protective measures are not taken to limit their movement towards the structure. A maintenance vessel used for dredging in the intake bay could also impact the intake structure. The screening of each source that could initiate an impact should be performed and the screened-in sources should be listed.

Detailed evaluation

9.11. Hazard analysis of screened-in sources should be performed to check the interaction with the nuclear installation. If there is an interaction, load characterization is required to be performed (see Requirement 7 and para. 4.19 of SSR-1 [1]) by considering a ship or barge moving with a conservatively estimated velocity.

9.12. In broad terms, the evaluation process should consider a distressed or incorrectly navigated vessel impacting a submerged, offshore or coastal structure of a nuclear installation. The evaluation of impacts depends on the number of vessel movements per year by size and inventory, the location of shipping lanes in relation to the location of the structure, and the ability to accurately model how a distressed vessel might come to impact such a structure. These aspects should be considered in the evaluation process.

9.13. Once the potential for impact has been established, the energy of impact should be calculated and other load characterization parameters estimated. Although in principle there are similarities between vessel impacts with marine structures and other types of projectile impact addressed in this Safety Guide, the nature of vessels (i.e. high mass, low speed) and the type of structures being considered may be quite different, and this should be taken into account.

Hazard parameters

9.14. The following are examples of hazard parameters that should be considered (see Table 2 in the Appendix):

(a) Passage routes (e.g. seaways) and frequency of passage;
(b) Frequency, type and route of movements to and from the source of HIEEs;
(c) Existing protective measures on passage routes.

Load characterization parameters

9.15. The following are examples of load characterization parameters that should be considered (see Table 4 in the Appendix):

(a) Impact energy at the shoreside of the nuclear installation or at an offshore facility location:
— Mass;
— Velocity;
— Size, cross-section area of plane of impact, and penetrative capability.
(b) Type of missile: soft missile.
(c) Direction of approach.

CARGO AND PIPELINES CONTAINING HAZARDOUS MATERIAL

9.16. The hazards associated with the surface transport of hazardous material include hazardous liquids and gases released on the ground (see Section 5), explosions (see Section 6) and fire (see Section 7). The same methodology should be used as for the mobile sources of HIEEs by taking the minimum distance from the nuclear installation site. Hazardous liquids discharged into seas and rivers are also addressed in this section.

9.17. Major pipelines in the region of the site should be evaluated, as they may carry hazardous liquids and gases. Such pipelines can leak from valves or as a result of an accident, and these should be considered.

9.18. An important route for interaction with the nuclear installation is provided by the water intake; a hazard could arise from a spillage at an adjacent installation or from a tanker accident (e.g. after uncontrolled drifting). Parameters for the dilution and dispersion of the liquid and its entry into the water intake should be evaluated. Consideration should be given to the fact that the spillage of highly flammable liquids on water can produce floating pools, which might approach a nuclear installation on the shore or along a riverbank. A conservative estimate should be made, and dispersion characteristics should be considered. Consideration should also be given to the possibility that liquids with low flash points might be extracted from contaminated sources of intake water. Other cargo that is not formally classified as hazardous material, such as thick liquids, pastes, absorbent bulky freight (e.g. wood pellets) and sticky chemicals, should also be considered in terms of its ability to jeopardize the water intake.

9.19. Liquids discharged from marine and river vessels disperse in response to local tide and/or river current conditions and can be carried several kilometres from the release point. For liquids released into a large body of water, dilution can be anticipated as the distance from the release point and the elapsed time increase, but the rate of dilution can be highly dependent on the local tide and current flow conditions at the time of release. Modelling of the way discharges are dispersed should be carried out. Alternatively, it can be assumed conservatively that no dilution occurs.

HAZARD ASSESSMENT FOR HUMAN INDUCED EXTERNAL EVENTS INVOLVING CARGO AND PIPELINES CONTAINING HAZARDOUS MATERIAL

Identification of sources of HIEEs

9.20. Sources of HIEEs involving hazardous liquids and gases are listed in Table 2 in the Appendix. Recommendations on data collection are provided in Section 4. First, the regions of interest should be located on the basis of generic screening distance values (see Table A–1 in the Annex). Sources within these regions constitute the hazardous material being transported; information (e.g. on types and quantities of hazardous material, frequency, routes) should be available from relevant local or national government agencies with responsibility for controlling access to transport routes. Data on potential sources of HIEEs should be collected and the distances between these sources and the nuclear installation site should be calculated.

Screening using distance

9.21. Simple calculations of screening distance can be made using the source data, and specific screening distance values should be estimated for the largest spills of hazardous material considered possible, assuming conservative parameters for dispersion and local tide and current flow conditions at the time of release. Those sources that lie farther away from the nuclear installation site can be screened out.

Screening using probability

9.22. If a hazard cannot be screened out by distance, generic event data (i.e. based on the total occurrence frequency of an event category) can be used. Pragmatic and conservative judgement should be applied to determine the probability of potential events involving the spillage of hazardous material. If the total probability of occurrence is less than the screening probability level, it can be screened out.

9.23. If the potential hazard from screened-in sources is likely to be less than that from similar materials stored on the nuclear installation site itself and against which protection has already been provided (i.e. protection that is also effective against hazards from off-site sources), it can be screened out. If several sources are screened out on the same basis, it may be necessary to reflect the frequency contribution arising from the sum of all such sources by nominating a bounding source and screening on this basis. The screening of each event that can affect the nuclear installation site from spillage in the sea or a river should be completed, and the screened-in sources should be listed.

Detailed evaluation

9.24. Hazard analysis of screened-in sources should be performed to check the interaction with the nuclear installation. If there is an interaction, load characterization is required to be performed (see Requirement 7 and para. 4.19 of SSR-1 [1]). Materials released into the sea or a river could disperse and dilute in complex ways that need explicit modelling by experts to determine the different types of hazardous material travel in the sea or river and how these might affect the structures or equipment of the nuclear installation, and to calculate the load characterization parameters.

Hazard parameters

9.25. The following are examples of hazard parameters that should be considered for load characterization:

(a)   The location of the transport route around the closest approach to the nuclear installation site;
(b)   The nature and quantities of hazardous material transported and in spillages;
(c)   Meteorological and hydrological conditions;
(d)   Relevant bathymetric, tidal and river current conditions around this route that might influence the dispersion and hazardous characteristics of a release.

Load characterization parameters

9.26. The following are examples of load characterization parameters that should be considered:

(a)   Concentration of hazardous material in cooling water at the intake;
(b)   The impact on a once through cooling water system.

10. OTHER HUMAN INDUCED EXTERNAL EVENTS

10.1. This section deals with those HIEEs that are not addressed in Sections 5–9. The hazards arising from these HIEEs are listed in Table 3 in the Appendix. Some regions surrounding a nuclear installation site may contain other hazards; however, it is not possible to comprehensively identify all possible hazards in this Safety Guide.

GROUND SUBSIDENCE HAZARDS FROM HUMAN INDUCED EXTERNAL EVENTS

10.2. The ground at a nuclear installation site can subside owing to a local geotechnical issue under the site or outside the site area due to human-made features such as mines, exploitation of natural gas fields, water wells and oil wells.

10.3. Paragraph 5.29 of SSR-1 [1] states:

"The potential for collapse, subsidence or uplift of the surface that could affect the safety of the nuclear installation over its lifetime shall be evaluated using a detailed description of subsurface conditions obtained from reliable methods of investigation."

All geotechnical and geological issues that could exclude a nuclear installation site should be evaluated during the site selection stage. Recommendations on geotechnical issues are provided in IAEA Safety Standards Series No. NS-G-3.6, Geotechnical Aspects of Site Evaluation and Foundations for Nuclear Power Plants [6], and recommendations on geological issues are provided in IAEA Safety Standards Series No. SSG-9 (Rev. 1), Seismic Hazards in Site Evaluation for Nuclear Installations [2].

10.4. For existing sites, whenever new construction work is planned on or near the site, subsidence issues should be studied, especially where deep excavation work is planned (e.g. for nuclear power plants). The issue is more complicated when nuclear installations are built on saturated soft soils with a high water table and dewatering is necessary. In such cases, it should be verified that dewatering does not lead to unacceptable (differential) settlement of the existing nuclear installation, and this should be monitored. Reinjection of the extracted water may be necessary to keep pore pressures at the existing nuclear installation unaltered during dewatering and the restoration period thereafter.

10.5. Large scale mining activities, exploitation of natural gas fields and extraction of oil and groundwater in the vicinity of the site can lead to subsidence. A specific assessment should be conducted in such cases, and no screening distance value can be provided as it will depend on the magnitude of the mining or oil or groundwater extraction activities and distance from the nuclear installation site.

Detailed evaluation

10.6. Engineering solutions are available to handle the subsidence from local effects but depend on the type of work to be undertaken and might not always be feasible. Engineering solutions to counter subsidence from HIEEs should be implemented after a detailed evaluation; such solutions might not be possible but administrative measures (e.g. restrictions on mining and the exploitation of natural gas fields, water wells and oil wells) in the site vicinity might be available. As such, a decision to select a site should be taken after a detailed evaluation.

Hazard parameters

10.7. The following are examples of hazard parameters that should be considered (see Table 2 in the Appendix):

(a)   Location and nature of adjacent groundworks;
(b)   Location and nature of underground works;
(c)   Relevant geological, hydrogeological and geotechnical ground conditions;
(d)   Details of planned activities in the site vicinity (e.g. mining, oil and water extraction).

Load characterization parameters

10.8. The following are examples of load characterization parameters that should be considered (see Appendix) if a site can be selected:

(a)   Settlement, differential settlement and settlement rate;
(b)   Existing engineered mitigation measures (for existing sites) or anticipated engineered measures (for new sites).

ELECTROMAGNETIC INTERFERENCE HAZARDS FROM HUMAN INDUCED EXTERNAL EVENTS

10.9. Electromagnetic interference can affect the functionality of electronic devices. It can be initiated by both on-site sources of electromagnetic radiation (e.g. high voltage switchgear, portable telephones, portable electronic devices, computers) and off-site sources (e.g. radio transmitters, military radar stations, particle accelerators, high voltage transmission lines, telephone networks). Particular attention should be paid to any jamming facilities used by the on-site security organization or by transmitters operated by national security authorities (e.g. airborne, seaborne or ground-located on the site or off the site), as information might not be available on the actual power and antenna amplification of these transmissions and the electromagnetic radiation power of the transmissions might be increased significantly with little or no warning. When information cannot be obtained, the regulatory body should be asked to estimate the significance of these hazards.

10.10. The process of identification of potential sources of electromagnetic interference and quantification should be continued during the lifetime of the nuclear installation to ensure proper protection of plant components, as the greater use of digital equipment in instrumentation and control systems is increasing the vulnerability to electromagnetic interference.

10.11. Generic screening distance values have not been developed by States for electromagnetic interference; therefore, it should be managed on a site specific basis for each nuclear installation site.

Detailed evaluation

10.12. A detailed evaluation is required to be conducted to establish the hazard parameters and load characterization (see Requirement 7 and para. 4.19 of SSR-1 [1]).

10.13. The electromagnetic fields at the point of installation for instrumentation and control systems that are important to safety should be assessed to identify any unique electromagnetic radiation sources that could generate local interference. The sources could include both portable and fixed equipment (e.g. portable transceivers, arc welding equipment, power supplies, generators).

Hazard parameters

10.14.   The following are examples of hazard parameters that should be considered (see Table 2 in the Appendix):

(a)   Frequency band and energy of emissions of electromagnetic radiation from sources at and around the site;

(b)   Existing protective measures at the source locations.

Load characterization parameters

10.15.   The following are examples of load characterization parameters that should be considered (see Table 4 (10) in the Appendix):

(a)   Frequency band and energy rating of protective measures against electromagnetic interference;

(b)   Existing engineered mitigation measures (existing sites).

HAZARDS DUE TO HUMAN INDUCED EXTERNAL EVENTS AT BOMBING AND FIRING RANGES

10.16.   This hazard should be handled in a special way if the bombing and firing ranges are within the generic screening distance value (see Table A–1 in the Annex). As information is not easily available for military sites, efforts should be made through government channels to obtain the necessary information about activities on the bombing and firing ranges[11]. The history of events and incidents outside the designated area for bombing and firing practice should be collected and used in the assessment. Information on the frequency of hung ordnance[12], flight path(s) taken to a recovery site, and frequency of dropped ordnance should be collected. A confidentiality agreement may need to be signed to not disclose

---

[11] If there are undisclosed national security locations (e.g. permanent underwater minefields, electronic warfare installations, concealed munitions depots) near the site that might cause a hazard for the nuclear installation, the operating organization of the installation or the regulatory body should make their best efforts to contact the responsible authorities to determine and minimize the potential hazard to the installation.

[12] Military flights carrying ordnance to a bombing range may encounter hung ordnance while discharging and recover by flying to a recovery airport/airfield (site). Hung ordnance are those weapons or stores on an aircraft that the pilot has attempted to drop or fire but could not because of a malfunction of the weapon, rack or launcher or aircraft release and control system.

any information. Any screened-in hazards are required to be evaluated (see Requirement 7 and para. 4.19 of SSR-1 [1]).

HAZARDS DUE TO MISCELLANEOUS HUMAN INDUCED EXTERNAL EVENTS

10.17.   The following events that might occur in the vicinity of the site should also be considered:

(a)   A severe accident at nearby nuclear installations (radiation hazard). Detailed guidance on the studies and investigations necessary for assessing the impact of a nuclear installation on humans and the environment is provided in NS-G-3.2 [5];
(b)   Disturbances in the connection of an external electric grid, including its unavailability;
(c)   Damage to headrace or tailrace facilities (in the case of once through cooling water on river sites).

11. EVALUATION OF EXTERNAL HUMAN INDUCED HAZARDS FOR NUCLEAR INSTALLATIONS OTHER THAN NUCLEAR POWER PLANTS

11.1. A graded approach is required to be applied to the evaluation of HIEEs on the basis of the complexity of the nuclear installation and the potential radiological hazards and other hazards (see paras 4.1 and 4.4 of SSR-1 [1]). This approach may be applied for each HIEE separately.

11.2. Prior to categorizing a nuclear installation for the purpose of applying a graded approach (see paras 11.9–11.12), a conservative screening process should be applied in which it is assumed that the entire radioactive inventory of the installation is released by an accident initiated by an HIEE. If the potential result of such a radioactive release were that no unacceptable consequences would be likely for workers, the public or the environment, and provided that no other specific requirements are imposed by the regulatory body for such an installation, no further HIEE hazard assessment needs to be performed.

11.3. If the results of the conservative screening process show that the potential consequences of such a release would not be acceptable, the hazards associated with HIEEs are required to be evaluated (see Requirement 7 and para. 4.19 of SSR-1 [1]).

11.4. The likelihood that an HIEE will give rise to radiological consequences will depend on the characteristics of the nuclear installation (e.g. its purpose, layout, design, construction and operation). Paragraph 4.5 of SSR-1 [1] states:

"For site evaluation for nuclear installations other than nuclear power plants, the following shall be taken into consideration in the application of a graded approach:

(a) The amount, type and status of the radioactive inventory at the site (e.g. whether the radioactive material on the site is in solid, liquid and/or gaseous form, and whether the radioactive material is being processed in the nuclear installation or is being stored on the site);
(b) The intrinsic hazards associated with the physical and chemical processes that take place at the nuclear installation;
(c) For research reactors, the thermal power;
(d) The distribution and location of radioactive sources in the nuclear installation;
(e) The configuration and layout of installations designed for experiments, and how these might change in future;
(f) The need for active systems and/or operator actions for the prevention of accidents and for the mitigation of the consequences of accidents;
(g) The potential for on-site and off-site consequences in the event of an accident."

11.5. Other factors that should be taken into account in the application of a graded approach include the following:

(a) The characteristics of engineered safety features for the prevention of accidents and for mitigation of the consequences of accidents (e.g. the containment and containment systems);
(b) The characteristics of the processes or the engineering features that might show a cliff edge effect in the event of an accident;
(c) The characteristics of the site relevant to the consequences of the dispersion of radioactive materials in the atmosphere and in the hydrosphere (e.g. size and demographics of the region).

11.6. Some nuclear installations may be located below the surface. Most HIEEs are expected to have limited potential to affect the safety of a subsurface installation, although those that can induce ground failure or affect ventilation systems should be considered. Any effects will depend on the HIEEs the installation is subjected to and the nature of the installation. Services supplied to subsurface installations could also be affected by HIEEs.

11.7. Other criteria may be specified by the regulatory body in relation to the application of a graded approach. For example, fuel damage, a radioactive release or radiation exposure may be the conditions or metrics of interest.

11.8. The application of a graded approach should be based on the following information:

(a)   The current safety analysis report for the installation (if available), which should be the primary source of information;
(b)   The results of an HIEE hazard assessment, if one has been performed;
(c)   The characteristics of the installation listed in paras 11.4 and 11.5.

11.9. The application of a graded approach should be based on a categorization of the installation. This may have been performed at the design stage or later. In general, the criteria for categorization should be based on the radiological consequences of the release of radioactivity from the installation, ranging from very low radiological consequences to potentially severe radiological consequences. Alternatively, the categorization may range from radiological consequences within the installation itself, to radiological consequences confined to the site boundary of the installation, to radiological consequences to the public and the environment outside the site.

11.10.   Three or more categories of nuclear installation may be defined on the basis of national practice and criteria. As an example, the following categories may be defined:

(a)   The lowest hazard category includes those nuclear installations for which national building codes for conventional facilities (e.g. essential facilities such as hospitals) or for hazardous facilities (e.g. petrochemical plants, chemical plants), at a minimum, should be applied.
(b)   The highest hazard category contains installations for which standards and codes for nuclear power plants should be applied.
(c)   There is often at least one intermediate category of hazardous installation, for which, at a minimum, codes dedicated to hazardous facilities should be

used. The number of intermediate categories will depend on the nature of the installation and whether the site contains a single or multiple nuclear installations or units.

11.11. In applying a graded approach to nuclear installations, it should be noted that installations other than nuclear power plants might not have sufficient inherent robustness against HIEEs. It might also be excessively costly to protect them against some HIEEs through design (e.g. the crash of a large aircraft). For new nuclear installations, necessary precautions should be taken at an early stage to protect the nuclear installation through appropriate siting whereby ample screening distance values are provided for major HIEEs.

11.12. The HIEE hazard evaluation should be performed in accordance with the following recommendations:

(a) For installations in the lowest hazard category (e.g. zero power research reactors), the HIEE hazard evaluation may be based on national building codes and standards, as established for important facilities within the State.

(b) For installations in the highest hazard category, the HIEE hazard evaluation should be implemented in the same manner as for nuclear power plants.

(c) For installations categorized in the intermediate hazard category (e.g. research reactors with a low to medium power), the following may be applicable:

(i) If the HIEE hazard evaluation is performed using methods similar to those described in this Safety Guide, a lower HIEE hazard level (i.e. than for nuclear power plants) for designing these installations may be adopted at the design stage, in accordance with the design requirements for the installation.

(ii) If the database and the methods recommended in this Safety Guide are found to be disproportionately complex, time consuming and demanding for the nuclear installation in question, simplified methods for HIEE hazard evaluation may be used. In such cases, the hazard parameter finally adopted for designing the installation should be commensurate with the reduced database and the simplification of the methods, with account being taken of the fact that both factors tend to increase uncertainties.

12. APPLICATION OF THE MANAGEMENT SYSTEM TO THE EVALUATION OF HUMAN INDUCED EXTERNAL EVENTS

12.1. Requirement 2 of SSR-1 states that **"Site evaluation shall be conducted in a comprehensive, systematic, planned and documented manner in accordance with a management system."**

12.2. A management system is required to be established, applied and maintained in accordance with IAEA Safety Standards Series No. GSR Part 2, Leadership and Management for Safety [20]. It should be applied for the activities performed in relation to the evaluation of hazards associated with HIEEs in site evaluation for nuclear installations.

ASPECTS OF PROJECT MANAGEMENT FOR THE EVALUATION OF HUMAN INDUCED EXTERNAL EVENTS

12.3. A project work plan for addressing HIEEs should be established that, at a minimum, addresses the following topics:

(a) The objectives and scope of the project;
(b) Applicable regulations and standards;
(c) Organization of the roles and responsibilities for management of the project;
(d) Work breakdown, processes and tasks, schedule and milestones;
(e) Interfaces among the different types of task (e.g. data collection tasks, analysis tasks) and disciplines involved, especially the various specialists needed for the different types of HIEE encountered with all necessary inputs and outputs;
(f) Project deliverables and reporting procedures.

12.4. The project scope should identify all the hazards generated by HIEEs that are relevant to the safety of the nuclear installation and that will be investigated within the framework of the project (see also Requirement 3 of SSR-1 [1]). If some HIEEs are not included within the scope, a justification should be provided.

12.5. The project work plan should include a description of all requirements that are relevant for the project, including applicable regulatory requirements in relation to all the hazards considered to be within the project scope. The applicability of these regulatory requirements should be reviewed by the regulatory body before the operating organization conducts the HIEE hazard evaluation.

12.6. All approaches and methodologies that reference lower tier legislation (e.g. regulatory guidance documents, industry codes and standards) should be clearly identified and described. If experts are consulted to better capture epistemic uncertainties, the sophistication and complexity of these approaches should be chosen by the study sponsor on the basis of the project requirements. The details of the approaches and methodologies to be used should be clearly stated in the project work plan.

12.7. The following management system process should be applied to ensure the quality of the project (see IAEA Safety Standards Series No. GS-G-3.1, Application of the Management System for Facilities and Activities [21]):

(a) Document control;
(b) Control of products;
(c) Control of measuring and testing equipment;
(d) Control of records;
(e) Control of analyses;
(f) Purchasing (procurement);
(g) Validation and verification of software;
(h) Audits (e.g. self-assessment, independent assessments, review);
(i) Control of non-conformances, corrective actions and preventive actions;
(j) Processes covering field investigations, laboratory testing, data collection, and analysis and evaluation of observed data;
(k) Communication processes for the interaction among the experts involved in the project.

12.8. The project work plan should ensure that there is adequate provision, in the resources and in the schedule, for collecting new data and/or analyses that might be important for the conduct of the HIEE hazard evaluation. This may arise, for example, where potential HIEEs have been identified at sources where the level of detail in the associated safety analysis is appropriate for the industry the source is associated with but is insufficient for the evaluation of hazards for a nuclear installation.

12.9. To ensure that the evaluation of hazards associated with HIEEs is traceable and transparent to users (e.g. peer reviewers, the operating organization, the regulatory body, the designers, the vendors, the contractors and the subcontractors of the operating organization), the documentation should provide a description of all elements of the evaluation process and include the following information:

(a) A description of the study participants and their roles;
(b) Background material including the data collection and analysis process and documentation, as well as the source display map;
(c) A description of the computer software used, and the input and output files;
(d) Reference documents;
(e) All documents supporting the treatment of uncertainties, opinion and related discussions;
(f) Results of intermediate calculations and sensitivity studies.

This documentation should be maintained in an accessible, usable and auditable form by the operating organization.

12.10. The documentation and references should identify all sources of information used in the HIEE hazard evaluation, including information on where to find important citations and other information that might be difficult to obtain. Unpublished data used in the analysis should be included in the documentation in an appropriately accessible and usable form. Where data that are restricted for security or commercial reasons have been used (see para. 4.1), it may be necessary to prepare redacted versions of documents. However, where such documents are used or passed to others (e.g. by peer reviewers or nuclear installation designers) as part of the HIEE hazard evaluation, the project organization should be responsible for ensuring that sufficient information is provided to enable tasks to be performed effectively and in the best interests of nuclear safety.

ENGINEERING USES AND OUTPUT SPECIFICATION IN THE EVALUATION OF HUMAN INDUCED EXTERNAL EVENTS

12.11. In addition to site evaluation, an HIEE hazard evaluation is usually conducted for the purposes of design and/or safety assessment of the nuclear installation. Therefore, the work plan for the HIEE hazard evaluation should identify the intended engineering uses and objectives of the evaluation and should incorporate an output specification that describes all the results necessary for the intended engineering uses and objectives of the study (see also para. 4.1).

INDEPENDENT PEER REVIEW OF THE EVALUATION OF HUMAN INDUCED EXTERNAL EVENTS

12.12. Paragraph 3.4 of SSR-1 [1] states that "The results of studies and investigations conducted as part of the site evaluation shall be documented in sufficient detail to permit an independent review."

12.13. Paragraph 3.5 of SSR-1 [1] states:

"An independent review shall be made of the evaluation of the natural and human induced external hazards and the site specific design parameters, and of the evaluation of the potential radiological impact of the nuclear installation on people and the environment."

12.14. An independent peer review should be conducted and implemented to provide assurance that (i) a proper process has been duly followed in conducting the HIEE hazard evaluation, (ii) the analysis has addressed and evaluated the involved uncertainties (both epistemic and aleatory), and (iii) the documentation is complete and traceable.

12.15. The members of the independent peer review team should have the necessary multidisciplinary expertise to address all technical and process related aspects of the HIEE hazard evaluation. The peer reviewers should not have been involved in other aspects of the project and should not have a vested interest in the outcome. The level and type of peer review can differ, depending on the application of the HIEE hazard evaluation.

12.16. One of the following two methods of peer review should be used: participatory peer review or late stage peer review. A participatory peer review is carried out during the HIEE evaluation process, allowing the reviewers to resolve comments. A late stage peer review is carried out toward the end of the evaluation. Participatory peer review will decrease the likelihood of the assessment's being found unsuitable at a late stage.

Appendix

TABLES TO BE USED IN THE EVALUATION OF HAZARDS ASSOCIATED WITH HUMAN INDUCED EXTERNAL EVENTS

A.1. The tables in this appendix provide information for use in evaluating hazards associated with HIEEs. Table 1 lists the categories of HIEEs, and Tables 2–4 provide information on their identification, evolution and possible effects, and potential impact on nuclear installations.

TABLE 1. CATEGORIES OF HUMAN INDUCED EXTERNAL EVENTS

Category of human induced external event	Generic screening distance values (see Table A–1 in the Annex)
(a) *External release of hazardous material.* This includes radioactive and other hazardous gases and liquids, pressurized and liquefied gases and flammable gases and liquids.	(1) (2) (3) (4)
(b) *External explosions.* These can arise from operational installations and/or stores containing explosive materials and/or undertaking processes with such materials that create situations where an enhanced potential for explosions exists.	(1) (2) (4)
(c) *External fire.*	(1) (3)
(d) *Aircraft crash.* This includes the categorization of different types of aircraft for hazard evaluation purposes, the characterization of aircraft movements near to a site and the modelling of an aircraft crash event so that the hazard can be parameterized and quantified. Air corridors should also be included when characterizing aircraft movements.	(5)

TABLE 1. CATEGORIES OF HUMAN INDUCED EXTERNAL EVENTS (cont.)

Category of human induced external event	Generic screening distance values (see Table A–1 in the Annex)
(e) *External transport events excluding aircraft crashes*. These can arise from road and rail vehicles, pipelines, river barges and sea vessels. Hazards in this category normally arise directly from crash events, which can lead to the release of hazardous gases, and fire and explosion events.	(1) (2) (3) (4)
(f) *Other HIEEs*. These include hazards arising from stationary and mobile sources not included in (a)–(e). This includes subsidence, electromagnetic interference and bombing and firing ranges.	Not applicable to subsidence and electromagnetic interference; (6) for bombing and firing ranges

TABLE 2. IDENTIFICATION OF SOURCES OF HUMAN INDUCED EXTERNAL EVENTS, EVENT CATEGORIES, HUMAN INDUCED EXTERNAL EVENTS AND SOURCE RELATED INFORMATION

Types of source		Category of event	HIEEs	Relevant source related information to be collected
Stationary sources				
(1)	Oil refineries, chemical plants, storage depots, broadcasting networks, mining or quarrying operations, dams and dock facilities, peat and forests, other nuclear installations, underground gas storage, fracking, ground works adjacent to the nuclear installation site	(a) External release of hazardous material	— Release of flammable, explosive, asphyxiant, corrosive or toxic material — Radioactive release from nearby nuclear facilities	— Quantity and nature of all materials and physical properties, chemistry, radiochemistry, flashpoint, toxicity or definition of other hazardous effects — Detailed information of nearby nuclear facilities (e.g. type, power) — Maximum credible release, or frequency versus quantity release relationship — Meteorological and topographical characteristics of the region — Below ground flows — geological seepage and flow routes and opportunities for material concentration — Existing protective measures at the source location (e.g. bunds)

TABLE 2. IDENTIFICATION OF SOURCES OF HUMAN INDUCED EXTERNAL EVENTS, EVENT CATEGORIES, HUMAN INDUCED EXTERNAL EVENTS AND SOURCE RELATED INFORMATION (cont.)

Types of source	Category of event	HIEEs	Relevant source related information to be collected
			— Parameters allowing the determination of the release rate of the flammable source (e.g. evaporation rate in the case of a flammable pool of hydrocarbon and release rate for flammable gas release) — Types and features of nuclear facilities
	(b) External explosions	— Deflagration wave (over-pressurization) — Detonation wave — Boiling liquid expanding vapour explosion — Exothermic chemical reaction — Dust explosion	— Nature of explosive material — Maximum credible pressure (over/under) and thermal release at source location, or explosion frequency versus severity relationship — Meteorological and topographical characteristics of the region — Existing protective measures at the source location (e.g. blast walls)

TABLE 2. IDENTIFICATION OF SOURCES OF HUMAN INDUCED EXTERNAL EVENTS, EVENT CATEGORIES, HUMAN INDUCED EXTERNAL EVENTS AND SOURCE RELATED INFORMATION (cont.)

Types of source		Category of event	HIEEs	Relevant source related information to be collected
(c)	External fire		— Hydrocarbon fire — Chemical fires other than hydrocarbon	— Nature of flammable material (e.g. soot, toxic products) and thermal release — Flashpoint, flammability concentrations in air or other ignition criteria — Maximum credible material or thermal release, or fire frequency versus severity relationship — Meteorological and topographical characteristics of the region — Existing protective measures at the source location (e.g. fire breaks)
(d)	Aircraft crash		— See (3)	
(e)	External transport events excluding aircraft crashes		— See (4)	— See (4)(e) — Frequency, type and route of movements to and from the source

71

TABLE 2. IDENTIFICATION OF SOURCES OF HUMAN INDUCED EXTERNAL EVENTS, EVENT CATEGORIES, HUMAN INDUCED EXTERNAL EVENTS AND SOURCE RELATED INFORMATION (cont.)

Types of source		Category of event	HIEEs	Relevant source related information to be collected
	(f)	Other HIEEs	— Projectiles and missiles — Ground subsidence — Electromagnetic interference — Bombing and firing ranges — Miscellaneous HIEEs	— Nature of projectile or missile (e.g. mass, initial velocity, trajectory) — Maximum credible projectile or missile, or frequency of release — Location and nature of adjacent ground works — Location and nature of underground works — Meteorological and topographical characteristics of the region — Relevant geological, hydrogeological and geotechnical ground conditions — Frequency band and energy of electromagnetic emissions — Existing protective measures against electromagnetic interference at the source location — Details of mining and fracking
(2)	Military facilities (permanent and temporary)	(a) External release of hazardous material	— Release of flammable, explosive, asphyxiant, corrosive, toxic or radioactive material	— See (1)(a).

TABLE 2. IDENTIFICATION OF SOURCES OF HUMAN INDUCED EXTERNAL EVENTS, EVENT CATEGORIES, HUMAN INDUCED EXTERNAL EVENTS AND SOURCE RELATED INFORMATION (cont.)

Types of source	Category of event	HIEEs	Relevant source related information to be collected
	(b) External explosions	— Deflagration — Detonation — Dust explosion	— See (1)(b)
	(c) External fire	— Hydrocarbon fire — Chemical fire	— See (1)(c)
	(d) Aircraft crash	— See (3)	— See (3)(d) — Frequency, type and route of movements to and from the source
	(e) External transport events excluding aircraft crashes	— See (4)	— See (4)(e) — Frequency, type and route of movements to and from the source
	(f) Other HIEEs	— Projectiles and missiles — Electromagnetic interference — Bombing and firing ranges	— See (1)(f)

TABLE 2. IDENTIFICATION OF SOURCES OF HUMAN INDUCED EXTERNAL EVENTS, EVENT CATEGORIES,
HUMAN INDUCED EXTERNAL EVENTS AND SOURCE RELATED INFORMATION (cont.)

Types of source		Category of event	HIEEs	Relevant source related information to be collected
Mobile sources				
(3) Airport facilities, air traffic	(a)	External release of hazardous material	— Release of flammable, explosive, asphyxiant, corrosive, toxic or radioactive material	— See (1)(a)
	(b)	External explosions	— Deflagration — Detonation	— See (1)(b)
	(c)	External fire	— Hydrocarbon fire	— See (1)(c)
	(d)	Aircraft crash	— Initiating events not covered in (3)(a, b, c, f) — Crash related to take-off and landing — Other sources of aircraft crash (e.g. background crash rate, airways)	— Information not covered in (3)(a, b, c, f) — Types and characteristics of aircraft — Aircraft movements and flight frequencies from airports — Runway orientation, length and location

TABLE 2. IDENTIFICATION OF SOURCES OF HUMAN INDUCED EXTERNAL EVENTS, EVENT CATEGORIES, HUMAN INDUCED EXTERNAL EVENTS AND SOURCE RELATED INFORMATION (cont.)

Types of source	Category of event		Relevant source related information to be collected
			— Airfield plates for take-off, landing and manoeuvring — Traffic type and frequencies in airways — Location, elevations and cross-section characteristics of airways — Location and characteristics of restricted, controlled and other forms of airspace — Types and characteristics of aircraft (e.g. mass, fuel load, speeds for various stages of flight) — National and regional crash data
(e)	External transport events excluding aircraft crashes	— See (4)	
(f)	Other HIEEs	— Projectiles, missiles and drones	— See (1)(f)

TABLE 2. IDENTIFICATION OF SOURCES OF HUMAN INDUCED EXTERNAL EVENTS, EVENT CATEGORIES, HUMAN INDUCED EXTERNAL EVENTS AND SOURCE RELATED INFORMATION (cont.)

Types of source		Category of event	HIEEs	Relevant source related information to be collected
(4) Railway trains and wagons, road vehicles, ships, barges, pipelines	(a)	External release of hazardous material	— Release of flammable, explosive, asphyxiant, corrosive, toxic or radioactive material — Blockage, contamination (such as from an oil spill) or damage to cooling water intake structures	— See (1)(a) — Location of transport routes and the closest approach to the nuclear installation site — Relevant topographic features in the region around these routes that might influence the dispersion and hazardous characteristics of a release — Relevant bathymetric, tidal and river current conditions around this route that might influence the dispersion and hazardous characteristics of a release
	(b)	External explosions	— Deflagration — Detonation	— See (1)(b) — Tidal and bathymetric characteristics of the region
	(c)	External fire	— Hydrocarbon fire — Chemical fire	— See (1)(c) — Tidal and bathymetric characteristics of offshore and near-shore region
	(d)	Aircraft crash	— See (3)	

TABLE 2. IDENTIFICATION OF SOURCES OF HUMAN INDUCED EXTERNAL EVENTS, EVENT CATEGORIES, HUMAN INDUCED EXTERNAL EVENTS AND SOURCE RELATED INFORMATION (cont.)

Types of source	Category of event	HIEEs	Relevant source related information to be collected
(e)	External transport events excluding aircraft crashes	— Initiating events not covered in (4)(a, b, c, f) — Vehicle or vessel impact — Vehicle derailment, or misdirection — Leak of hazardous material from a pipeline	— Information not covered in (4)(a, b, c, f) — Passage routes and frequency of passage (e.g. road and rail routes, seaways) — Location and routing of pipelines and associated pumping stations — Frequency, type and route of movements to and from the source — Existing protective measures for vehicles, vessels and routes — Transportation accidents data
(f)	Other HIEEs	— Projectiles and missiles — Electromagnetic interference	— See (1)(f)

77

TABLE 3. EVOLUTION OF SOURCES OF HUMAN INDUCED EXTERNAL EVENTS AND POSSIBLE EFFECTS ON NUCLEAR INSTALLATIONS

Event category	HIEEs	Possible hazard at the nuclear installation site	Possible effects on the nuclear installation (see Table 4)
(a) External release of hazardous material	— Release of flammable, explosive, asphyxiant, corrosive or toxic material — Release of radioactivity from nearby nuclear facilities — Explosion — Hydrocarbon fire — Other types of chemical fire — Projectiles and missiles	— Clouds or liquids can drift toward the nuclear installation and burn or explode before or after reaching it, outside or inside the installation — Clouds or liquids can also migrate into areas and affect operating personnel or items important to safety — Radiation exposures to operating personnel at the nuclear installation	(5) (6) (8)
(b) External explosions	— Deflagration — Detonation — Dust explosion — Release of flammable, explosive, asphyxiant, corrosive, toxic or radioactive material	— Explosion pressure wave — Projectiles — Smoke, gas and dust produced in explosion can drift toward the nuclear installation	(1) (2) (3) (4) (7) (8)

TABLE 3. EVOLUTION OF SOURCES OF HUMAN INDUCED EXTERNAL EVENTS AND POSSIBLE EFFECTS ON NUCLEAR INSTALLATIONS (cont.)

Event category	HIEEs	Possible hazard at the nuclear installation site	Possible effects on the nuclear installation (see Table 4)
(c) External fire	— Hydrocarbon fire — Chemical fires other than hydrocarbon — Projectiles and missiles		
	— Hydrocarbon fire — Chemical fires other than hydrocarbon — Explosion — Release of flammable, explosive, asphyxiant, corrosive, toxic or radioactive material — Projectiles and missiles	— Associated flames and fires; sparks can ignite other fires — Smoke and combustion gas can drift towards the nuclear installation — Heat (thermal flux)	(3) (4) (5)

(d) Aircraft crash	— Crash related to take-off and landing
	— Other sources of aircraft crash:
	  • Background crash rate, airways
		— Primary effects:
		  • Impact damage to structures including perforation, penetration
		  • Vibration effects
			(1) (2) (3) (4) (6)

TABLE 3. EVOLUTION OF SOURCES OF HUMAN INDUCED EXTERNAL EVENTS AND POSSIBLE EFFECTS ON NUCLEAR INSTALLATIONS (cont.)

Event category	HIEEs	Possible hazard at the nuclear installation site	Possible effects on the nuclear installation (see Table 4)
	• Release of flammable, explosive, corrosive, toxic or radioactive material • Explosion • Hydrocarbon fire • Missiles	• Global stability — Secondary effects: 　• Secondary missiles ejected from the impact site and scattering widely • Rapid spread of flammable liquid from the point of impact, including impulsive damage to structures from the momentum of the released liquid when ejected from the aircraft • Fire and explosion generating heat and blast effects and generating tertiary missiles	

TABLE 3. EVOLUTION OF SOURCES OF HUMAN INDUCED EXTERNAL EVENTS AND POSSIBLE EFFECTS ON NUCLEAR INSTALLATIONS (cont.)

Event category	HIEEs	Possible hazard at the nuclear installation site	Possible effects on the nuclear installation (see Table 4)
		• Release of hazardous material carried as cargo • Ground shaking	
(e) External transport events excluding aircraft crashes	— Vehicle impact — Vehicle derailment, or misdirection — Release of flammable, explosive, asphyxiant, corrosive, toxic or radioactive material — Blockage, contamination (such as from an oil spill) or damage to cooling water intake structures — Explosion — Hydrocarbon fire — Chemical fires other than hydrocarbon — Projectiles and missiles	— Direct impact damage — Secondary projectiles — Fire — Explosion of fuel tanks or cargo	(2) (4) (7) (8) (11)

TABLE 3. EVOLUTION OF SOURCES OF HUMAN INDUCED EXTERNAL EVENTS AND POSSIBLE EFFECTS
ON NUCLEAR INSTALLATIONS (cont.)

Event category	HIEEs	Possible hazard at the nuclear installation site	Possible effects on the nuclear installation (see Table 4)
(f) Other HIEEs	— Projectiles and missiles — Subsidence — Electromagnetic interference — Release of large volumes of water or change of watercourse — Bombing and firing ranges	— Missile impact with structure — Ground failure under structures — Flooding onto the nuclear site, or change of water table — Direct damage to structures and equipment — Fire as secondary effect — Electromagnetic fields around electrical equipment leading to failure, malfunction or spurious electrical signals	(2) (7) (8) (9) (10) (11)

TABLE 4. IMPACT ON THE NUCLEAR INSTALLATION AND CONSEQUENCES

Possible hazard effects on the nuclear installation	Load characterization parameters	Consequences of hazard effects
(1) Pressure wave	— Local overpressure at the installation as a function of time	— Damage or collapse of parts of structure or disruption of systems and components — Secondary hazards (e.g. fire, explosion, release of hazardous material)
(2) Projectile	— Impact energy at nuclear installation location (mass, velocity) — Compass direction and angle of approach from horizontal — Missile hardness and penetrative capability in structures important to safety (e.g. shape, size, type of material) — Existing protective measures at the source location	— Damage to structures (e.g. penetration, perforation, spalling, scabbing, collapse of structures) — Disruption or failure of structures, systems and components including buried systems and services — Induced vibration — Loss of access or egress for emergency and/or safety related operator actions — Secondary hazards (e.g. fire, explosion, release of hazardous material)
(3) Heat	— Maximum temperature flux and duration	— Impaired habitability of control room — Disruption of systems or components — Damage to structures — Ignition of combustibles — Secondary effects (e.g. sparks, fires, smoke)

83

TABLE 4. IMPACT ON THE NUCLEAR INSTALLATION AND CONSEQUENCES (cont.)

Possible hazard effects on the nuclear installation	Load characterization parameters	Consequences of hazard effects
(4) Smoke and dust	— Composition — Concentration and quantity as a function of time	— Blockage of ventilation intake filters and diesel engine air filters — Impaired habitability of control room and other areas important to the safety of the nuclear installation
(5) Asphyxiant and toxic material	— Concentration and quantity as a function of time — Volatility in ambient conditions — Toxicity and asphyxiant limits	— Threat to operating personnel and impaired habitability of the main control room and other areas important to the safety of the nuclear installation — Incapacitation of operating personnel or reduced ability to perform safety related tasks
(6) Corrosive and radioactive liquids, gases and aerosols	— Concentration and quantity as a function of time — Corrosive, radioactive limits — Provenance (sea, land)	— Threat to operating personnel and impaired habitability of areas important to the safety of the nuclear installation — Corrosion and disruption of systems or components, loss of strength — Electrical short circuits — Blockage of water intakes, site drains — Prevention of fulfilment of safety functions
(7) Ground shaking	— Frequency response spectrum for vibrational motion	— Mechanical damage

TABLE 4. IMPACT ON THE NUCLEAR INSTALLATION AND CONSEQUENCES (cont.)

Possible hazard effects on the nuclear installation	Load characterization parameters	Consequences of hazard effects
(8) Flooding or drought	— Elevation of site above main water course or mean sea level — Level of water with time — Velocity of impacting water	— Damage to structures, systems and components due to inundation — Damage to structures, systems and components directly or functional failure due to water impact — Damage to structures, systems and components or functional failure due to secondary effects such electrical short circuit — Loss of safety functions requiring water (in case of drought)
(9) Subsidence	— Settlement, differential settlement, settlement rate — Existing engineered mitigation measures (existing sites), or anticipated measures (new sites)	— Collapse of structures, disruption or failure of structures, systems and components including buried systems and services — Secondary hazards (e.g. fire, explosion, release of hazardous material)
(10) Electromagnetic interference	— Frequency band and energy rating of protection against electromagnetic interference — Existing engineered mitigation measures (existing sites), or anticipated measures (new sites)	— Incorrect or spurious electrical signals in items important to safety leading to spurious operation or action
(11) Damage to water intake	— Mass of the ship, lost cargo, impact velocity and area, degree of blockage	— Unavailability of cooling water

REFERENCES

[1] INTERNATIONAL ATOMIC ENERGY AGENCY, Site Evaluation for Nuclear Installations, IAEA Safety Standards Series No. SSR-1, IAEA, Vienna (2019).

[2] INTERNATIONAL ATOMIC ENERGY AGENCY, Seismic Hazards in Site Evaluation for Nuclear Installations, IAEA Safety Standards Series No. SSG-9 (Rev. 1), IAEA, Vienna (2022).

[3] INTERNATIONAL ATOMIC ENERGY AGENCY, WORLD METEOROLOGICAL ORGANIZATION, Meteorological and Hydrological Hazards in Site Evaluation for Nuclear Installations, IAEA Safety Standards Series No. SSG-18, IAEA, Vienna (2011).

[4] INTERNATIONAL ATOMIC ENERGY AGENCY, Volcanic Hazards in Site Evaluation for Nuclear Installations, IAEA Safety Standards Series No. SSG-21, IAEA, Vienna (2012).

[5] INTERNATIONAL ATOMIC ENERGY AGENCY, Dispersion of Radioactive Material in Air and Water and Consideration of Population Distribution in Site Evaluation for Nuclear Power Plants, IAEA Safety Standards Series No. NS-G-3.2, IAEA, Vienna (2002). (A revision of this publication is in preparation.)

[6] INTERNATIONAL ATOMIC ENERGY AGENCY, Geotechnical Aspects of Site Evaluation and Foundations for Nuclear Power Plants, IAEA Safety Standards Series No. NS-G-3.6, IAEA, Vienna (2004). (A revision of this publication is in preparation.)

[7] INTERNATIONAL ATOMIC ENERGY AGENCY, Design of Nuclear Installations Against External Events Excluding Earthquakes, IAEA Safety Standards Series No. SSG-68, IAEA, Vienna (2021).

[8] INTERNATIONAL ATOMIC ENERGY AGENCY, Protection Against Internal and External Hazards in the Operation of Nuclear Power Plants, IAEA Safety Standards Series No. SSG-77, IAEA, Vienna (2022).

[9] INTERNATIONAL ATOMIC ENERGY AGENCY, IAEA Nuclear Safety and Glossary: Terminology Used in Nuclear Safety, Nuclear Security, Radiation Protection and Emergency Preparedness and Response, 2022 (Interim) Edition, IAEA, Vienna (2022).

[10] INTERNATIONAL ATOMIC ENERGY AGENCY, Nuclear Security Recommendations on Physical Protection of Nuclear Material and Nuclear Facilities (INFCIRC/225/ Revision 5), IAEA Nuclear Security Series No. 13, IAEA, Vienna (2011).

[11] INTERNATIONAL ATOMIC ENERGY AGENCY, Physical Protection of Nuclear Material and Nuclear Facilities (Implementation of INFCIRC/225/Revision 5), IAEA Nuclear Security Series No. 27, IAEA, Vienna (2018).

[12] INTERNATIONAL ATOMIC ENERGY AGENCY, Security During the Lifetime of a Nuclear Facility, IAEA Nuclear Security Series No. 35-G, IAEA, Vienna (2019).

[13] INTERNATIONAL ATOMIC ENERGY AGENCY, National Nuclear Security Threat Assessment, Design Basis Threats and Representative Threat Statements, IAEA Nuclear Security Series No. 10-G (Rev. 1), IAEA, Vienna (2021).

[14] INTERNATIONAL ATOMIC ENERGY AGENCY, Engineering Safety Aspects of the Protection of Nuclear Power Plants Against Sabotage, IAEA Nuclear Security Series No. 4, IAEA, Vienna (2007).

[15] INTERNATIONAL ATOMIC ENERGY AGENCY, Identification of Vital Areas at Nuclear Facilities, IAEA Nuclear Security Series No. 16, IAEA, Vienna (2012)

[16] INTERNATIONAL ATOMIC ENERGY AGENCY, Safety of Research Reactors, IAEA Safety Standards Series No. SSR-3, IAEA, Vienna (2016).

[17] INTERNATIONAL ATOMIC ENERGY AGENCY, Safety of Nuclear Fuel Cycle Facilities, IAEA Safety Standards Series No. SSR-4, IAEA, Vienna (2017).

[18] INTERNATIONAL ATOMIC ENERGY AGENCY, Safety Aspects of Nuclear Power Plants in Human Induced External Events: Assessment of Structures, Safety Reports Series No. 87, IAEA, Vienna (2018).

[19] INTERNATIONAL ATOMIC ENERGY AGENCY, Safety Aspects of Nuclear Power Plants in Human Induced External Events: General Considerations, Safety Reports Series No. 86, IAEA, Vienna (2017).

[20] INTERNATIONAL ATOMIC ENERGY AGENCY, Leadership and Management for Safety, IAEA Safety Standards Series No. GSR Part 2, IAEA, Vienna (2016).

[21] INTERNATIONAL ATOMIC ENERGY AGENCY, Application of the Management System for Facilities and Activities, IAEA Safety Standards Series No. GS-G-3.1, IAEA, Vienna (2006). (A revision of this publication is in preparation.)

Annex

TYPICAL GENERIC SCREENING DISTANCE VALUES

A–1. Table A–1 provides typical generic screening distance values used by some States for large nuclear power plants with standardized designs. These generic screening values can be used as a basis for identifying source regions centred on a nuclear installation site. Generic screening distance values are intended to be conservative. When using these values, it needs to be ensured that they are appropriate for the HIEEs likely to occur at each source considered.

TABLE A–1. TYPICAL GENERIC SCREENING DISTANCE VALUES

Sources		Generic screening distance value
(1)	Facilities for storing or handling flammable, corrosive or explosive material	5–10 km
(2)	Sources of hazardous clouds, vapours or gases	8–10 km
(3)	Sources of fire such as forests, peat, storage areas for low volatility flammable materials (especially hydrocarbon storage tanks), wood or plastics, factories that produce or store such materials, their transport lines, and vegetation	1–2 km
(4)	Military installations storing munitions	8 km
(5)	Aircraft crash events:	
	(i) An aircraft crash at the site resulting from the general air traffic in the region (Type 1 aircraft crash)	Not applicable, see para. 8.11
	(ii) An aircraft crash at the site resulting from take-off or landing manoeuvres at a nearby airport (Type 2 aircraft crash)	8 km
	(iii) An aircraft crash at the site resulting from air traffic in the main civil traffic corridors and military flight zones (Type 3 aircraft crash)	4 km
(6)	Distance from military installations or air space usage such as bombing and firing ranges	30 km

CONTRIBUTORS TO DRAFTING AND REVIEW

Altinyollar, A.	International Atomic Energy Agency
Blahoianu, A.	Consultant, Canada
Coman, O.	International Atomic Energy Agency
Contri P.	International Atomic Energy Agency
Ford, P.	Consultant, United Kingdom
Guerpinar, A.	Consultant, Türkiye
Guinard, L.	Électricité de France, France
Guner, B.	Nuclear Regulatory Authority, Türkiye
Henkel, F.O.	Consultant, Germany
Mahmood, H.	Consultant, Pakistan
Morita, S.	Nuclear Energy Agency
Orbovic, N.	Canadian Nuclear Safety Commission, Canada
Preece, A.	Office for Nuclear Regulation, United Kingdom

 **IAEA**
International Atomic Energy Agency

ORDERING LOCALLY

IAEA priced publications may be purchased from the sources listed below or from major local booksellers.

Orders for unpriced publications should be made directly to the IAEA. The contact details are given at the end of this list.

NORTH AMERICA

Bernan / Rowman & Littlefield
15250 NBN Way, Blue Ridge Summit, PA 17214, USA
Telephone: +1 800 462 6420 • Fax: +1 800 338 4550
Email: orders@rowman.com • Web site: www.rowman.com/bernan

REST OF WORLD

Please contact your preferred local supplier, or our lead distributor:

Eurospan Group
Gray's Inn House
127 Clerkenwell Road
London EC1R 5DB
United Kingdom

Trade orders and enquiries:
Telephone: +44 (0)176 760 4972 • Fax: +44 (0)176 760 1640
Email: eurospan@turpin-distribution.com

Individual orders:
www.eurospanbookstore.com/iaea

For further information:
Telephone: +44 (0)207 240 0856 • Fax: +44 (0)207 379 0609
Email: info@eurospangroup.com • Web site: www.eurospangroup.com

Orders for both priced and unpriced publications may be addressed directly to:

Marketing and Sales Unit
International Atomic Energy Agency
Vienna International Centre, PO Box 100, 1400 Vienna, Austria
Telephone: +43 1 2600 22529 or 22530 • Fax: +43 1 26007 22529
Email: sales.publications@iaea.org • Web site: www.iaea.org/publications